RAND NATIONAL DEFENSE RESEARCH INSTITUTE

T0288867

NATO's Amphibious Forces

Command and Control of a Multibrigade Alliance Task Force

Gene Germanovich, J.D. Williams, Stacie L. Pettyjohn,
David A. Shlapak, Anthony Atler, Bradley Martin

Prepared for the United States Marine Forces Europe and Africa

For more information on this publication, visit www.rand.org/t/RR2928

Library of Congress Cataloging-in-Publication Data is available for this publication.
ISBN: 978-1-9774-0236-3

Published by the RAND Corporation, Santa Monica, Calif.
© Copyright 2019 RAND Corporation
RAND® is a registered trademark.

Cover: Photo by Robert L. Kunzig , NATOChannel.

Support RAND
Make a tax-deductible charitable contribution at
www.rand.org/giving/contribute

www.rand.org

Preface

In 2017 and 2018, the RAND Corporation designed and facilitated a series of wargames and seminars on behalf of the Amphibious Leaders Expeditionary Symposium (ALES), a forum for general and flag officers to discuss opportunities for improved interoperability, command and control (C2), and utilization of amphibious forces within the North Atlantic Treaty Organization (NATO). With a scenario centered on confrontation with a near-peer competitor providing context, maritime and amphibious leaders explored how to leverage the Alliance's existing amphibious capacity by aggregating national capabilities under a coherent C2 structure.

This report provides a summary of RAND-facilitated ALES events, the methodology employed, and the resultant observations and recommendations. We also highlight several overarching findings from the research to inform NATO's evolving command structure, upcoming exercises, and planning for collective defense.

This research was sponsored by U.S. Marine Corps Forces Europe and Africa and conducted within the International Security and Defense Policy Center of the RAND National Defense Research Institute, a federally funded research and development center sponsored by the Office of the Secretary of Defense, the Joint Staff, the Unified Combatant Commands, the U.S. Navy, the U.S. Marine Corps, the defense agencies, and the intelligence community.

For more information on the RAND International Security and Defense Policy Center, see www.rand.org/nsrd/ndri/centers/isdp or contact the director (contact information is provided on the webpage).

Contents

Figures and Tables

Summary

The amphibious capabilities of the North Atlantic Treaty Organization (NATO) include five European nations—France, Italy, the Netherlands, Spain, and the United Kingdom—with large L-class amphibious ships and corresponding landing forces, as well as the uniquely large and global force provided by the U.S. Marine Corps and U.S. Navy. Amphibious leaders from these nations and Portugal, whose marines routinely train and operate alongside Spanish marines, participate in the Amphibious Leaders Expeditionary Symposium (ALES), a forum for general and flag officers to discuss opportunities for improved interoperability, command and control (C2), and utilization of amphibious forces within NATO.[1] Meeting since 2016, ALES participants have focused on integrating their forces to contribute to NATO's deterrence posture and collective defense at the major joint operation plus (MJO+) level.[2]

In 2017–2018, U.S. Marine Corps Forces Europe and Africa asked the RAND Corporation to design and facilitate three events with the objective of identifying suitable C2 constructs and associated doctrine, organization, training, materiel, leadership, personnel, facilities, and interoperability (DOTMLPF-I) considerations for large-scale maritime and amphibious operations in support of NATO. Aided by a scenario centered on confrontation with a near-peer competitor, maritime and amphibious leaders explored how to leverage the Alliance's existing amphibious capacity by aggregating national capabilities into a coherent C2 structure.

- **The Naples Tabletop.** In June 2017, maritime and amphibious leaders from ALES nations and NATO's joint and maritime commands participated in a tabletop exercise (TTX) in Naples, Italy. The TTX explored challenges with NATO's current maritime C2 arrangements and identified implications for allied amphibious operations in contested environments.

---

[1] Other NATO members with amphibious elements in their armed forces include Belgium, Germany, Greece, Romania, and Turkey. Several others maintain coastal and special operations-focused forces that have more limited amphibious capabilities.

[2] The term *MJO+* denotes a large-scale allied military campaign.

- **The Northwood High North Seminar.** In November 2017, amphibious leaders met in Northwood, United Kingdom, for a facilitated seminar that advanced the discussion from the Naples Tabletop and proposed a candidate amphibious C2 construct—referred to as the centralized amphibious task force (ATF)—for additional wargaming.
- **The Stavanger Wargame.** In June 2018, maritime and amphibious leaders reconvened in Stavanger, Norway, to explore the utility and function of the centralized ATF construct, better understand its viability in an MJO+ scenario, and examine the necessary actions NATO requires to plan, test, and implement the construct.

Research Question and Method

The overarching question of this research effort was, "How could NATO organize its amphibious forces for a conventional conflict against a near-peer competitor?" Answering this question required deriving answers to the following subordinate questions:

- Which national or NATO commands, if any, have the requisite capabilities to oversee large-scale, multinational maritime operations that include amphibious assault?
- Who could serve as the commander, amphibious task force (CATF) and commander, landing force (CLF) of a NATO ATF?
- What DOTMLPF-I enhancements would be needed to fully realize ambitions for a NATO ATF?

Participants at ALES events included flag and general officers from eight allied nations and eight NATO organizations. Table S.1 lists participating nations and organizations, including U.S. commands. The group's expertise and seniority enabled a realistic discussion on how the Alliance could arrange C2 for the maritime and amphibious aspects of an MJO+ if one occurred.

The unclassified and fictional scenario used for wargaming and seminar discussion featured a conflict with allies invoking Article 5 of the NATO Treaty and Supreme Allied Commander Europe (SACEUR) ordering an amphibious assault to retake territory in the North Atlantic region. Wargaming did not attempt to test the viability of amphibious operations in any existing plans, but rather served as a backdrop for discussions regarding C2 and interoperability. Each event used variations of wargaming and/or seminar-based expert elicitation techniques.

Participants at the Naples Tabletop assessed the current C2 construct for large-scale allied maritime operations as unsuitable for an MJO+ against a near-peer adversary. There was no consensus on the exact shape that the C2 construct should take, but all participants concluded that the existing construct was not appropriate for an

Table S.1
Participants at RAND-Facilitated ALES Events

Nations	NATO Commands and Organizations	U.S. Commands and Organizations
France	Allied Air Command	European Command
Italy	Allied Command Transformation	Africa Command
Netherlands	Joint Force Command Naples	Air Forces Europe and Africa
Norway	Allied Maritime Command (MARCOM)	Army Forces Europe
Portugal	Naval Striking and Support Forces NATO	Marine Corps Forces Europe and Africa
Spain	Joint Warfare Centre Norway	Naval Forces Europe and Africa
United Kingdom	Combined Joint Operations from the Sea Center of Excellence	Fleet Forces Command
United States	NATO International Military Staff	Marine Corps Forces Command
		II Marine Expeditionary Force

NOTE: Each entity listed sent at least one general or flag officer, or their representative, to participate in the Naples Tabletop, Northwood High North Seminar, and/or Stavanger Wargame.

MJO+. Participant recommendations and observations from Naples were forwarded to SACEUR and the Supreme Allied Commander Transformation for consideration during deliberations on the maritime aspects of NATO's Command Structure Adaptation process.[3]

Next, ALES turned to identifying an amphibious C2 concept that could "plug into" NATO's evolving maritime C2 structure. After reviewing several alternatives, amphibious leaders at the Northwood High North Seminar arrived at a baseline concept in the form of the centralized ATF. As envisioned, this organizational construct consists of a colocated CATF and CLF, along with supporting multinational staffs and subordinate naval and landing force components. Figure S.1, a simplified and representative C2 diagram, shows how this organization serves as an intermediate command between the combined force maritime component command (CFMCC) and smaller extant national and bilateral amphibious task groups (ATGs).[4] The ATF directs, coordinates, and synchronizes the ATGs as a coherent force.

---

[3] NATO Command Structure Adaptation is an initiative that includes the establishment of JFC Norfolk to oversee potential operations in the North Atlantic.

[4] Forces from the United Kingdom and the Netherlands maintain a habitual relationship and can operate as an integrated United Kingdom Netherlands Amphibious Force (UKNLAF). Spain and Italy have a similar relationship through the Spanish Italian Amphibious Force (SIAF) and Spanish Italian Landing Force (SILF) construct (SIAF/SILF). Spain may also embark Portuguese Marines on Spanish ships, in which case the ATG becomes trilateral. The United States and France do not maintain integrated amphibious forces with other nations.

Figure S.1
Simplified Centralized Amphibious Task Force Construct

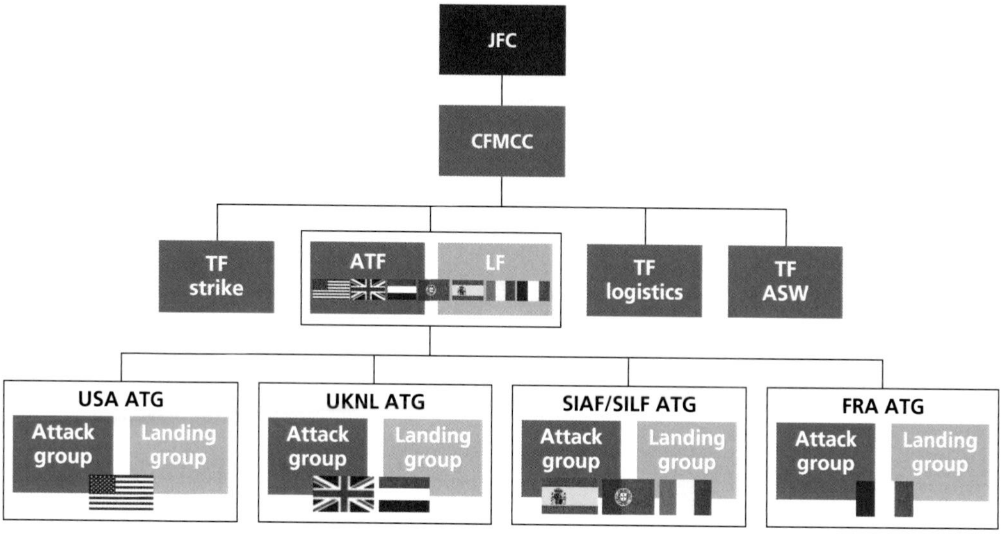

NOTE: ASW = anti-submarine warfare; JFC = joint force command; LF = landing force; FRA = France;
TF = task force. Participants designed the centralized ATF to "plug into" a broader maritime and joint
C2 architecture. The Naples Tabletop explored demands on the CFMCC, while subsequent events
focused on the composition and organization of the ATF and its subordinate ATGs.

Finally, the Stavanger Wargame explored the utility and function of the centralized ATF, how it could be employed in an MJO+, and what actions NATO requires to
develop this capability. The remainder of this summary provides findings based on the
totality of the three events and concludes by offering potential next steps to NATO.

Findings from the ALES Research

Finding 1: *ALES nations have considerable amphibious capacity, but these forces*
have been an underrecognized asset in NATO. Individual national and bilateral
ATGs currently form the core of allied amphibious forces. In the case of bilateral entities such as UKNLAF and SIAF/SILF, habitual training and deployments over the
course of decades have resulted in integral force packages with a level of uniformity
approaching that of purely national units such as the French and U.S. task groups.
These forces, while capable of conducting battalion- or brigade-size operations, most
frequently train and are employed below the battalion level. Further, current NATO
force planning and crisis response structures may not take advantage of the full potential of available amphibious capabilities.

As NATO responds to a changing security environment and moves to improve its maritime capabilities, amphibious forces offer a flexible and potent instrument that can enhance deterrence through early deployment and project credible combat power from the maritime domain. Allied ATGs maintain the requisite shipping, connectors, and landing forces to conduct multibrigade operations. U.S. forces, in the form of an expeditionary strike group (ESG) and Marine expeditionary brigade, are the most capable in penetrating anti-access/area denial environments, but European amphibious formations, when enabled by joint forces, could be employed in a range of scenarios including amphibious demonstrations, raids, subsidiary landings, and assaults against properly prepared objective areas. Energized by ALES, NATO is now considering how to transform the sum of these amphibious elements into something greater than its parts.

Finding 2: *The centralized ATF construct developed by ALES stakeholders offers a mechanism to leverage NATO's amphibious capacity by aggregating national and bilateral capabilities into a coherent C2 structure.* There is an emerging consensus around a baseline C2 structure—the centralized ATF—for amphibious operations in an MJO+ scenario. Maritime and amphibious leaders recognize that this construct requires testing and validation in exercises and is not appropriate for every situation. Wargaming enabled the discovery of three key challenges to the centralized ATF:

1. There is a need to delineate roles and responsibilities for a CFMCC and centralized ATF. The centralized ATF concept could enable a CFMCC to concentrate on the broader maritime campaign, but there are outstanding questions regarding prioritization, allocation, management, sequencing, and control of assets as amphibious operations unfold.

2. Even with CFMCC-ATF roles defined, employing a multinational ATF at echelons above a single brigade or battalion landing team poses new challenges for U.S. and allied amphibious forces. Complex aspects of battlespace management, such as integrated air and missile defense or cross-boundary fires integration, will require selection and refinement of C2 procedures that provide an appropriate blend of centralization and delegation of authorities while retaining operational flexibility in a potentially degraded communications environment.

3. In task organizing, commanders will need to balance the flexibility to tailor force groupings afforded under a centralized ATF against the stability and continuity provided by employing national and bilateral forces as integral elements. Participants at ALES events emphasized that national and bilateral ATGs are most effective when employed as integral units and expressed reservations about the political and tactical risk inherent in mixing and matching forces, particularly ground maneuver elements. Many amphibious leaders nevertheless recognize that a multinational ATF could encounter situations where it would need

to shift resources rapidly; this includes the possibility of assigning an ATG, or a subset of an ATG, in support or under the temporary control of an adjacent force it may not have trained with. Such relationships could be exceptionally difficult to arrange during a contingency, making rehearsals in peacetime essential.

Finding 3: *The centralized ATF construct requires commanders and staff with experience in multinational operations and expertise in amphibious warfare.* ALES events demonstrated that the commanders of the ATF will be required to exercise broad, complex C2 functions that include coordination with the CFMCC and effective oversight of multiple ATGs. Individuals filling the roles of CATF and CLF must have the requisite rank and experience to perform these functions. In addition, they should have multinational staffs of sufficient size and expertise demanded by the scope and scale of the amphibious force and its operations.

- For the CATF, there is no clear source of a navy officer and core staff. The U.S. ESG is a potential source for this capability but would require augmentation and training. The recent reconstitution of the U.S. 2nd Fleet may offer additional options.
- For the CLF role, the United States is best positioned to provide a two- or three-star officer and core staff from within II Marine Expeditionary Force.
- The size of a multinational staff could be as high as 500 personnel, depending on composition of the ATF and division of functions between the CFMCC and the ATF.
- Given the C2 and communications and information systems (CIS) functionality required and the potential size of the ATF staff, the choice of an ATF flagship is an important planning factor. While U.S. platforms have frequently served as flagships for larger NATO maritime elements, several European ships appear to have potential to serve in this capacity.

Allied staff planners could define manning requirements for the centralized ATF to inform deliberations about CATF and CLF selection, staff composition, flagship selection, and whether the organization should be standing, rotational, or sourced at time of incident.

Finding 4: *Knowledge and experience in large-scale amphibious operations has atrophied across allied naval and landing force practitioners.* Commanders and staffs should reexamine and reinforce maritime and amphibious doctrine for large-scale operations. Throughout ALES discussions, participants often had diverging perspectives on terminology and concepts. This may be a result of participants defaulting to concepts and terminology based on recent exercise and operational experiences rather than established NATO doctrine. Another possible explanation is that existing doctrine may not be sufficiently clear regarding its application to operations of a multinational force above the brigade level. Regardless, allies should stress the development of broader amphibious expertise in training and exercises as well as invest efforts in evolving doctrinal and operational concepts for the conduct of amphibious operations against new and emerging threats.

Potential Next Steps for NATO

The ALES focus on maritime and amphibious C2 has identified key issues, surfaced potential solutions, and stimulated action within national, bilateral, multinational, and Alliance organizations. Seminars and wargames provided allied military leaders and their staffs with a structured process through which to design and explore alternative C2 constructs for amphibious operations. Figure S.2 shows the progression of design, wargaming, and exercising needed to arrive at a construct optimized for large-scale NATO operations.

With the initial rounds of C2-centric wargaming at the operational level completed, the next step is command post exercises (CPXs) and live exercises that include a multinational ATF and its command ship.[5] Events such as CPXs enable the creation or refinement of templates, orders, and standard operating procedures that amphibious leaders acknowledge need attention for multi-ATG operations under a centralized ATF. Exercises could also permit testing issues that are difficult to wargame but directly impact C2, such as CIS architecture and performance.

Although exercises present the most immediate mechanism to improve operational proficiency, generating an allied ATF capability requires a broader set of actions. Realizing a NATO ATF will require that the Alliance's political commitment to

Figure S.2
Steps for Exploring Command and Control as of Mid-2018

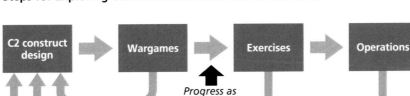

5 Exercises could be supplemented with tactical-level wargaming and modeling and simulation, which would be needed to explore the relative strengths and weaknesses of each ATG and inform expectations for most likely support relationships and reinforcement plans.

amphibious operations be reinforced with allied planning, national resourcing, and military implementation. The following six steps could form the core of a NATO strategy to evolve its amphibious capability:

1. Continue to develop NATO's maritime C2 structure to address MJO+ scope and complexity.
2. Exercise the envisioned amphibious C2 construct (centralized ATF) at the MJO+ scale.
3. Identify structure and candidate providers of multinational CATF/CLF and staff.
4. Draft an allied concept paper that describes what the centralized ATF is and how allies envision using it.
5. Develop a road map to operationalize the ATF capability.
6. Consider scaled-down versions of the ATF for a range of crisis response scenarios.[6]

Throughout wargames and seminars, participants noted that Allied Maritime Command (MARCOM), as SACEUR's maritime adviser, is uniquely positioned to work with ALES stakeholders (and other NATO entities) to pursue this or a similar agenda. For example, MARCOM could advise SACEUR on how amphibious forces could be integrated into NATO's Readiness Initiative or inform the revision of graduated response plans in close consultation with allies that possess amphibious capabilities.[7] Implementation of the Allied Maritime Posture initiative provides another venue for operationalizing the centralized ATF concept.[8] Figure S.3 summarizes ALES progress since 2016 and offers potential next steps for MARCOM and other NATO stakeholders.

Chapter Summary

ALES events surfaced the potential benefits of the centralized ATF while also pointing to actions needed for planning, testing, and implementing this C2 construct. Participants—many of today's maritime and amphibious leaders within the Alliance—came away with a clearer understanding of the challenges they face in preparing their forces

---

[6] This could include examining the role of amphibious forces in the NATO Response Force and Standing NATO Maritime Groups.

[7] The NATO Readiness Initiative aims to enhance the readiness of existing national forces and their ability to move within Europe and across the Atlantic; allies have committed, by 2020, to having 30 battalions, 30 air squadrons, and 30 naval combat vessels ready to use within 30 days.

[8] The Allied Maritime Posture initiative is designed to inform allied investment and capability development for maritime forces, including amphibious forces.

Figure S.3
ALES Progress and Potential Next Steps for NATO

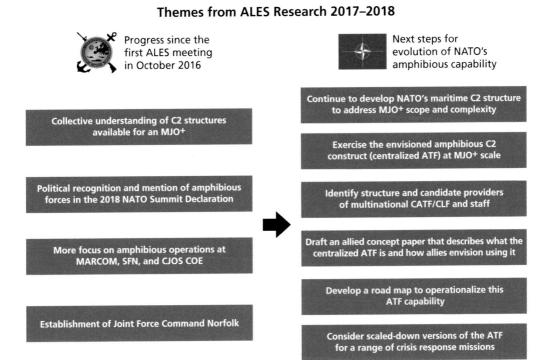

Themes from ALES Research 2017–2018

Progress since the first ALES meeting in October 2016

Next steps for evolution of NATO's amphibious capability

Collective understanding of C2 structures available for an MJO+

Political recognition and mention of amphibious forces in the 2018 NATO Summit Declaration

More focus on amphibious operations at MARCOM, SFN, and CJOS COE

Establishment of Joint Force Command Norfolk

Continue to develop NATO's maritime C2 structure to address MJO+ scope and complexity

Exercise the envisioned amphibious C2 construct (centralized ATF) at MJO+ scale

Identify structure and candidate providers of multinational CATF/CLF and staff

Draft an allied concept paper that describes what the centralized ATF is and how allies envision using it

Develop a road map to operationalize this ATF capability

Consider scaled-down versions of the ATF for a range of crisis response missions

NOTE: CJOS COE = Combined Joint Operations from the Sea Center of Excellence; SFN = Naval Striking and Support Forces NATO.

to integrate inside what is an evolving NATO command structure. More broadly, there is a recognition that while NATO has recently taken steps to enhance its preparedness for an MJO+, additional work remains in the amphibious realm to harness significant national and bilateral capacity into a coherent, integrated multinational capability in the form of a centralized ATF.

A final caveat is in order. While the wargame illustrated that the centralized ATF is an attractive option for organizing and employing NATO's amphibious forces, **no C2 construct can be successful without sufficiently ready and capable forces**. In this regard, properly resourcing amphibious forces and tailoring and sequencing a collective and progressive exercise schedule—to include national and bilateral events— are perhaps the most important considerations for NATO in building a potent and flexible amphibious capability for the Alliance.

Acknowledgments

The authors are grateful to the sponsors of this research, MajGen Niel E. Nelson, Commander, MARFOREUR/AF, from 2015 to 2017, and MajGen Russell A. C. Sanborn, Commander, MARFOREUR/AF, from 2017 to the time of this report's publication. Additionally, this work would not have been possible without the expertise and facilitation of the project monitors from MARFOREUR/AF's G-5 European Regional Plans staff. LtCol Gregory DeMarco's vision contributed directly to the June 2017 Naples Tabletop exercise's design and execution and laid the groundwork for future events. LtCol Lisa Deitle, Maj Justin Hooker, and Maj Kenneth Parisi ably guided the November 2017 Northwood High North Seminar and the June 2018 Stavanger Wargame. Aaron Daviet and David Kuehn provided advice throughout this research effort, LtCol Michael Meyer oversaw the project's conclusion, and Donald Paulson provided logistical support and analytic insights at each stage. Finally, numerous U.S. and allied officers contributed to what was a deeply collaborative effort across the maritime and amphibious community—there are too many to name, but everyone involved played key roles in event preparation and provided vital feedback on drafts of this report.

The authors also wish to recognize the contributions of the following RAND researchers and associates who were instrumental to the planning and execution of the research: Damien Baveye, Jeremy Boback, Michael Decker, Joslyn Fleming, Sarah Harting, Will Mackenzie, Igor Mikolic-Torreira, Meagan Smith, Tom Whitmore, Barry Wilson, and John Yurchak. CDR Colin Roberts and CDR T. J. Gilmore, the 2017 and 2018 U.S. Navy Fellows at RAND, were invaluable to the team, providing intimate knowledge of NATO and U.S. naval force capabilities.

Abbreviations

A2/AD	anti-access/area denial
ACT	Allied Command Transformation
AIRCOM	Allied Air Command
ALES	Amphibious Leaders Expeditionary Symposium
AO	area of operations
AOA	amphibious objective area
ARG/MEU	amphibious ready group/Marine expeditionary unit
ASW	antisubmarine warfare
ATF	amphibious task force
ATG	amphibious task group
BLT	battalion landing team
BMD	ballistic missile defense
C2	command and control
CAF	commander, amphibious force
CATF	commander, amphibious task force
CDCM	coastal defense cruise missile
CFACC	combined force air component command
CFLCC	combined force land component command
CFMCC	combined force maritime component command
CFMCC-NE	Combined Force Maritime Component Command–Northeast
CFSOCC	combined force special operations component command
CG	guided missile cruiser
CIS	communications and information systems
CJOS COE	Combined Joint Operations from the Sea Center of Excellence
CJSOR	comprehensive joint statement of requirements

CJTF	combined joint task force
CLF	commander, landing force
COMSUBNATO	Commander, NATO Submarine Forces
CONOPS	concept of operations
COP/CIP	common operating picture/common intelligence picture
CPX	command post exercise
CSG	carrier strike group
CV	aircraft carrier
CVN	aircraft carrier (nuclear)
CWC	composite warfare commander
DDG	guided missile destroyer
DOTMLPF-I	doctrine, organization, training, materiel, leadership, personnel, facilities, and interoperability
ESF	expeditionary strike force
ESG	expeditionary strike group
FFG	guided missile frigate
FFTA	frigate, towed array
FLT	fleet
HRF(M)	High Readiness Force (Maritime)
JFC	joint force command
JOA	joint operations area
JSS	joint strategic ship
LANDCOM	Allied Land Command
LCS	littoral combat ship
LF	landing force
LHA	amphibious assault ship
LHD	amphibious assault ship
LPD	landing platform dock
LSD	landing ship dock
LSD(A)	landing ship dock (auxiliary)
MARCOM	Allied Maritime Command
MARFOREUR/AF	U.S. Marine Corps Forces Europe and Africa
MCC	maritime component command
MCM	mine countermeasures

MEB	Marine expeditionary brigade
MJO+	major joint operation plus
MPA	maritime patrol aircraft
NAC	North Atlantic Council
NATO	North Atlantic Treaty Organization
OPCOM/OPCON	operational command/operational control
OTC	officer in tactical control
SACEUR	Supreme Allied Commander Europe
SFN	Naval Striking and Support Forces NATO
SIAF	Spanish Italian Amphibious Force
SILF	Spanish Italian Landing Force
SLOC	sea lines of communication
SSGN	nuclear guided missile submarine
SSK	attack submarine
SSN	nuclear-powered submarine
SUB	submarine force
TACOM/TACON	tactical command/tactical control
TF	task force
TG	task group
TTX	tabletop exercise
UKNLAF	United Kingdom Netherlands Amphibious Force
USW	undersea warfare

CHAPTER ONE

Introduction

The amphibious capabilities of the North Atlantic Treaty Organization (NATO) include amphibious ships and/or landing forces from Belgium, France, Greece, Germany, Italy, the Netherlands, Spain, Portugal, Turkey, and the United Kingdom, as well as the uniquely large and global force provided by the U.S. Marine Corps and U.S. Navy.[1] Amphibious leaders from the six nations with large L-class amphibious ships and corresponding landing forces (France, Italy, the Netherlands, Spain, the United Kingdom, and the United States) and Portugal, whose marines routinely train and operate alongside Spanish marines, participate in the Amphibious Leaders Expeditionary Symposium (ALES), a forum for general and flag officers to discuss opportunities for improved interoperability, command and control (C2), and utilization of amphibious forces within NATO. Meeting since 2016, ALES participants have focused on integrating their forces to contribute to NATO's deterrence posture and collective defense at the major joint operation plus (MJO+) level.[2]

In 2017, U.S. Marine Corps Forces Europe and Africa (MARFOREUR/AF) asked the RAND Corporation to design and facilitate a series of wargames and seminars to explore C2 constructs and concepts for a multinational amphibious task force (ATF) consisting of forces from ALES nations. This chapter summarizes each event's design and methods to enable discovery of C2 structures and enhancements that could better leverage NATO's amphibious capacity.

Research Question and Method

The overarching question of this research effort was, "How could NATO organize its amphibious forces for a conventional conflict against a near-peer competitor?" Answering this question required deriving answers to the following subordinate questions:

---

[1] Several other NATO nations maintain coastal and special operations-focused forces that have more limited amphibious capabilities.

[2] The term *MJO+* denotes a large-scale allied military campaign.

- Which national or NATO commands, if any, have the requisite capabilities to oversee large-scale, multinational maritime operations that include amphibious assault?
- Who could serve as the commander, amphibious task force (CATF) and commander, landing force (CLF) of a NATO ATF?
- What are the doctrine, organization, training, materiel, leadership, personnel, facilities, and interoperability (DOTMLPF-I) enhancements needed to fully realize ambitions for a NATO ATF?

With the realization that both national and collective amphibious knowledge had atrophied since the end of the Cold War, ALES participants identified the need to develop effective C2 arrangements as a prerequisite for integrating their forces at scale. They realized that a progression of design, wargames, and exercises were required to arrive at a construct optimized for large-scale NATO operations. Figure 1.1 shows the envisioned process. RAND's role was to facilitate exploration of the first two steps: design and wargaming. Two wargames and a seminar were planned in preparation for subsequent testing in allied exercises and potential inclusion in NATO's operational planning.

In coordination with MARFOREUR/AF and ALES stakeholders, RAND conducted three events with the objective of identifying suitable C2 constructs and associated DOTMLPF-I considerations in support of NATO maritime and amphibious capability evolution.

- **The Naples Tabletop.** In June 2017, maritime and amphibious leaders from ALES nations and NATO's joint and maritime commands participated in a tabletop exercise (TTX) in Naples, Italy. The TTX explored challenges with NATO's current maritime C2 arrangements and identified implications for allied amphibious operations in contested environments.
- **The Northwood High North Seminar.** In November 2017, amphibious leaders met in Northwood, United Kingdom, for a facilitated seminar that advanced the discussion from the Naples Tabletop and proposed a candidate amphibious C2

Figure 1.1
Steps for Exploring Command and Control

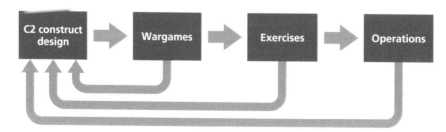

construct for additional wargaming. Amphibious leaders also discussed recent developments in the North Atlantic and Arctic security environment.

- **The Stavanger Wargame.** In June 2018, maritime and amphibious leaders reconvened in Stavanger, Norway, to explore the utility and function of the C2 construct as envisioned in Northwood, better understand its viability in an MJO+, and examine the necessary actions NATO requires to plan, test, and implement the construct.

The Naples Tabletop framed the overall C2 exploration by examining NATO's current command structure at the joint and maritime component levels while also beginning a discussion about alternatives for an amphibious C2 structure, which will need to "plug into" NATO's evolving command structure.[3] In Northwood and Stavanger, ALES participants focused on identifying plausible ATF C2 constructs for further evaluation and testing in command post and live exercises scheduled in the coming years. The range of maritime and amphibious C2 constructs considered appears in Appendix A.

The unclassified and fictional scenario used for wargaming and seminar discussion featured a conflict with allies invoking Article 5 of the NATO Treaty and Supreme Allied Commander Europe (SACEUR) ordering an amphibious assault to retake territory in the North Atlantic region. This MJO+ scenario, including the near-peer competitor's forces, is described in Appendix B. RAND made only minor adjustments to the scenario from event to event. Wargaming did not attempt to test the viability of amphibious operations, but rather served as a backdrop for discussions regarding C2 and interoperability. Each event used variations of wargaming and/or seminar-based expert elicitation techniques. Specific arrangements (i.e., number of game cells or seminar discussion groups, their assignments, and adjudication) are described for each event individually in Chapters Three through Five.

Participants

The core participants at each of the events were general and flag officers from ALES nations and NATO commands or organizations. Table 1.1 is a list of entities that participated in at least one event.

For each event, RAND worked in collaboration with MARFOREUR/AF and ALES stakeholders to ensure an appropriate balance across nationality, service, and rank. In terms of the latter, the senior attendee in Naples was a four-star admiral playing the role of commander for a joint force command (JFC); the Northwood discussion

---

[3] NATO is currently reforming its command structure, including the creation of JFC Norfolk to oversee potential operations in the North Atlantic region.

Table 1.1
Participants at RAND-Facilitated ALES Events

Nations	NATO Commands and Organizations	U.S. Commands and Organizations
France	Allied Air Command (AIRCOM)	European Command
Italy	Allied Command Transformation (ACT)	Africa Command
Netherlands	Joint Force Command Naples	Air Forces Europe and Africa
Norway	Allied Maritime Command (MARCOM)	Army Forces Europe
Portugal	Naval Striking and Support Forces NATO	Marine Corps Forces Europe and Africa
Spain	Joint Warfare Centre Norway	Naval Forces Europe and Africa
United Kingdom	Combined Joint Operations from the Sea Center of Excellence	Fleet Forces Command
United States	NATO International Military Staff	Marine Corps Forces Command
		II Marine Expeditionary Force

NOTE: Each entity listed sent at least one general or flag officer, or their representative, to participate in the Naples Tabletop, Northwood High North Seminar, and/or Stavanger Wargame.

was hosted by the three-star Commander of Allied Maritime Command and included one- and two-star generals participating in breakout discussion groups; and Stavanger game play was co-led by a three-star admiral and three-star general acting in the roles of the CATF and CLF. The group's expertise and seniority enabled a realistic discussion of how the Alliance could arrange C2 for the maritime and amphibious aspects of an MJO+. Indeed, if such a conflict occurred today, participants that attended these events would likely be the ones assigned to lead the maritime and amphibious aspects of NATO's response.

Limitations

A series of discussion-based events is insufficient to definitively determine the C2 constructs most suitable for operations. Although the maritime and amphibious community arrived at a concept for further testing in allied exercises, it should not be viewed as more than that: a candidate solution that may work depending on the situation and such factors as geopolitical context, enemy force capabilities, the disposition and state of friendly forces at time of crisis, and geography. Throughout the ALES series, participants were cognizant that no single C2 solution would likely accommodate the range of NATO's foreseeable amphibious operations.

This effort did not address the overall feasibility or advisability of conducting amphibious operations in the European theater—this would require extensive campaign planning coupled with modeling and simulation. Additionally, we did not attempt to replicate or simulate communications and information systems (CIS) or other technical aspects of C2. While some of these issues were raised, and are briefly noted in Appendix C, they would benefit from explorations and collaborative planning by ALES and NATO's amphibious community.

Finally, as discussed in Chapter Two, the joint and maritime portion of NATO's evolving command structure—prerequisites for amphibious operations—remains a work in progress. Appendix D identifies existing NATO organizations with the mission and ability to execute aspects of maritime C2, albeit with constraints identified at the Naples Tabletop.

Report Structure

This report is designed to meet two objectives: first, to provide findings to inform future national, ALES, and NATO preparations for maritime and amphibious exercises and operations in the context of collective defense against a capable aggressor; and second, to document the design and execution of the wargames and seminar. Chapter Two outlines the strategic context in which ALES and this research effort was undertaken. Chapters Three through Five detail the design, execution, and observations and recommendations of the Naples Tabletop, the Northwood High North Seminar, and the Stavanger Wargame, respectively. The concluding Chapter Six reviews progress in ALES's effort to arrive at C2 solutions, posits implications to NATO going forward, and offers some potential next steps. The appendixes include a depiction of the C2 constructs considered, the scenario and force list used for wargaming, DOTMLPF-I considerations discovered throughout research, and a brief overview of NATO's current options for fulfilling the role of a combined force maritime component command (CFMCC). The DOTMLPF-I considerations may be particularly useful to NATO planners as they develop requirements for future Alliance amphibious capabilities and consider additional force enhancements beyond C2 constructs.

> **Readers most interested in the dynamics that led to the ALES focus on amphibious C2 and the implications to NATO may elect to read Chapters Two and Six, and Appendix C.** The remaining text provides a detailed overview of each ALES event and how RAND employed wargaming and seminar facilitation to achieve research objectives.

CHAPTER TWO

Strategic Context

Prior to 2014, NATO's collective focus, and that of most individual allies, centered on preparedness for crisis response and security cooperation to project stability beyond its borders. Although NATO's 2010 *Strategic Concept* identified collective defense as a core task, the document assessed that "today, the Euro-Atlantic area is at peace and the threat of a conventional attack against NATO territory is low."[1] Allied militaries sought to conclude or downsize operations in Afghanistan and Iraq, rebalance to address a variety of threats such as global terrorism and piracy, and make the most of limited resources in a postrecession fiscal environment.

In this context, amphibious forces aimed to regenerate naval expertise eroded by over a decade of land-centric operations in the Middle East. For European nations the goal was to maintain and exercise the ships, connectors, and landing forces needed for operations with a battalion landing team (BLT) or less.[2] In terms of C2, planning focused on organizing national or bilateral amphibious task groups (ATGs) for situations short of MJO+.[3]

By 2018, however, the level of ambition grew to encompass a larger, multinational amphibious force construct to reinforce the Alliance's maritime posture and complement NATO's recent adaptations in the land domain. This chapter reviews the strategic context behind this shift and the set of issues MARFOREUR/AF asked RAND to consider while designing ALES wargames and a seminar in 2017–2018.

---

[1] NATO, *Strategic Concept for the Defence and Security of the Members of the North Atlantic Treaty Organization, Adopted by Heads of State and Government at the NATO Summit in Lisbon*, Brussels: North Atlantic Treaty Organization, November 19–20, 2010, p. 11.

[2] The U.S. Marine Corps, a much larger force than its European counterparts, was regenerating its proficiency to operate at the MEB level, but this, too, had a crisis response focus. See, for example, Donald Walton, "Dawn Blitz 2013: Training for Strength," NNS130128-01, Washington, D.C.: U.S. Navy, January 28, 2013.

[3] While there is no prescribed size for amphibious task elements, an ATG constituted for presence or small-scale contingency operations is generally understood to consist of two to five amphibious ships, while an ATF will normally have at least ten amphibious ships and numerous other types of vessels.

Interest in a Multinational Amphibious Capability (2014–2016)

Starting in 2014, three dynamics generated interest in allied maritime and amphibious C2 on a larger scale: an imperative for multinational interoperability and readiness resulting from trends in European security; momentum to create a multinational amphibious force by leveraging ALES and related forums; and concerns about NATO's existing maritime command structure and plans for employing an amphibious force during a major conflict.

The Imperative for Allied Interoperability and Readiness

Following Russia's 2014 annexation of Crimea, the subsequent conflict in eastern Ukraine, and continued instability to the Alliance's south, NATO members understood that no single nation has the capacity and political will to safeguard collective security on its own. Allies expressed an interest in increasing their force commitments, revisiting posture, and seeking to integrate forces to reassure allies and deter adversaries. Early adaptations included enhancing ground forces in the Baltic region and forming the Very High Readiness Joint Task Force as the spearhead for the NATO Response Force. At the 2016 NATO summit in Warsaw, allies also pledged to supplement these land-oriented actions by continuing "to reinforce our maritime posture by exploiting the full potential of the Alliance's overall maritime power."[4]

From the U.S. military's perspective, this meant that U.S. European Command and its maritime components needed to reexamine how naval power could be utilized in the European theater given existing U.S. Navy and Marine Corps global responsibilities. In the amphibious realm, MARFOREUR/AF identified an opportunity to combine U.S. maritime reach and capability with a sizable European capacity. As MARFOREUR/AF noted in its campaign plan,

> The collective amphibious capacity of European Allies and partners nearly matches that of the Navy–Marine Corps team world-wide. . . . In a crisis, non-U.S. NATO allies have the capacity to generate two to three brigades afloat on approximately 20 amphibious ships. MARFOREUR/AF will harness this capacity in order to generate sufficient European capability and readiness to deter aggression, complement U.S. forces in potential contingencies, and expand training opportunities in multinational settings.[5]

This plan signaled an increased U.S. interest in and contribution to ongoing efforts such as the European Amphibious Initiative to enhance amphibious interop-

---

[4]   NATO, "Warsaw Summit Communiqué," July 9, 2016a, item no. 48.

[5]   MARFOREUR/AF, "Roadmap for Maritime Expeditionary Operations: A Multinational Approach in Europe," in *Campaign Plan 2016–2020*, 2017, not available to the public. The quoted material is from the annex of the document, which the document owner has agreed to publication in this report.

erability and readiness in support of NATO and the European Union.[6] For NATO, enhancing U.S.-European maritime and amphibious collaboration has the potential of generating a larger, more capable force than currently available while sending a political signal that the Alliance is following through on its maritime commitments from Warsaw, which were later reinforced at the 2018 Brussels Summit.[7]

The Amphibious Leaders Expeditionary Symposium

Amphibious leaders from France, Italy, the Netherlands, Portugal, Spain, the United Kingdom, and the United States initiated discussion regarding a U.S.-European amphibious force at the inaugural ALES forum in October 2016. Seeking to capitalize on tactical interoperability developed in recent years,[8] ALES participants agreed to the following objective:

> Allied Marine forces will plan for a scalable, up to three-star Combined Landing Force capability—with organic ground, air, and logistics forces—in support of NATO. Participating nations will leverage, support, and integrate with existing NATO, European, and national exercises, initiatives, frameworks, and organizations.[9]

Given the inherent complexity of organizing and commanding even a single-nation amphibious operation, ALES discussion yielded a realization that establishing a viable C2 construct for the envisioned multinational force posed an urgent challenge. Questions such as "How should the multinational ATF be organized?" or "What is the appropriate maritime construct to command a multibrigade ATF?" had no immediate answers.

---

[6]   The European Amphibious Initiative's 2016 Declaration of Intent states that its purpose is

> to enhance European amphibious capability, primarily through establishing greater cooperation and progressively improving interoperability between existing forces. Developing an increasingly unified European amphibious capability will enhance the ability both of the European Union to act where NATO as a whole is not engaged, and will also improve NATO's effectiveness. (European Amphibious Initiative, presentation at the Amphibious Leaders Expeditionary Symposium, Stuttgart, Germany, October 2016)

[7]   NATO, "Brussels Summit Declaration," July 11, 2018, par. 19. Additionally, the Allied Maritime Posture initiative approved at this summit includes an amphibious component. Interview with MARFOREUR/AF staff, Washington, D.C., July 2018.

[8]   MARFOREUR/AF's Amphibious Maritime Basing Initiative involves a series of bilateral engagements to enhance tactical interoperability. Examples of recent initiatives include ship upgrades that enhance multinational communication and the certification of MV-22 Ospreys to operate aboard European ships. ALES leaders intend for this type of tactical interoperability to serve as a foundation for operational-level interoperability at the ATF level.

[9]   MARFOREUR/AF, "ALES 2016 After Action Report," Stuttgart: U.S. Marine Corps Forces Europe and Africa, 2016, p. 1.

Representatives of several NATO navies and NATO's maritime commands joined ALES discussions at another working group in early 2017, at which time a tentative maritime C2 construct, referred to as the "as is" arrangement, was identified for examination at the Naples Tabletop. ALES leaders also agreed to consider a challenging MJO+ scenario involving dozens of warships from the Alliance. The "as is" maritime C2 construct is depicted in Appendix A, while the notional list of U.S. and European maritime and amphibious forces available for employment by participants at the Naples Tabletop, and later the Stavanger Wargame, is provided in Appendix B.

NATO's Maritime Command Structure

NATO does not have Cold War–style standing plans that define command relationships, including maritime ones, for theater-wide operations. The JFCs, MARCOM, and Naval Striking and Support Forces NATO (SFN) all report directly to SACEUR in peacetime. Additionally, a High Readiness Force (Maritime), or HRF(M), headquarters is designated on an annual basis to serve as a CFMCC for the NATO Response Force. Appendix D identifies the roles and configurations of MARCOM, SFN, and the HRF(M) construct.[10]

In the event of a major conflict, North Atlantic Council (NAC) guidance will inform SACEUR's options for organizing forces and defining command relationships between the JFCs, MARCOM, SFN, and HRF(M) headquarters, and functional components for land, sea, air, and special operations forces. These arrangements will in turn inform CATF/CLF selection and amphibious C2 concept.

Although the Alliance has a series of graduated response plans, these are limited to the initial stages of a conflict and do not incorporate theater-wide operations on an MJO+ scale.[11] When examining this issue in 2016, ALES stakeholders recognized the need to refine C2 options and arrangements based on the emergence of a near-peer competitor. From the perspective of national forces, clarifying the roles and responsibilities of NATO's maritime commands would provide a sounder basis for planning.[12] The need to identify potential maritime and amphibious C2 solutions (along with key

---

[10] The Alliance is currently engaged in adapting its command structure. A new JFC is being created to oversee maritime operations in the North Atlantic, while the other maritime commands may receive updated missions and resources. See, for example, Jens Stoltenberg, "Press Conference," Brussels: North Atlantic Treaty Organization, June 7, 2018.

[11] Jens Ringsmose and Sten Rynning, "Now for the Hard Part: NATO's Strategic Adaptation to Russia," *Survival*, Vol. 59, No. 3, June–July 2017, p. 134.

[12] During the Cold War, NATO and allied navies delineated expected wartime C2 arrangements to offer the NAC a starting point if a maritime conflict with the Soviet Union was to occur. Indeed, a substantially larger NATO Command Structure was designed to account for a maritime campaign in the North Atlantic aimed at securing the sea lines of communication (SLOCs) for movement of U.S. land forces to continental Europe. Allies sought to achieve a level of automaticity in their plans to enable a rapid response in a politically feasible, operationally effective manner. For an extensive discussion of this topic, see Robert S. Jordan, *Alliance Strategy and Navies*, London: Pinter Publishers, 1990.

considerations for implementation across the DOTMLPF-I spectrum) served as the motivation for organizing the wargames and seminar.

Recent Developments (2017–2018)

In addition to C2-focused events covered in this report, ALES stakeholders also convened in January 2018 for a broader workshop focusing on the role of amphibious forces within NATO.[13] Three themes emerged from the discussion that relate to this research and ongoing effort to design, wargame, and exercise C2 for NATO's amphibious forces.

First, several naval experts noted that NATO's challenges in its maritime environs are manifold and often relevant to what amphibious forces offer. For one of the regions discussed at the workshop, the North Atlantic, the panelists and audience participants cautioned that while amphibious forces have promise, Russia's increasing anti-access/area denial (A2/AD) capabilities pose a major challenge. From a C2 perspective, this necessitates careful examination of command authorities for tasking, reallocation, and employment of maritime (and, in some cases, joint) assets for functions such as anti-submarine warfare (ASW) and air defense. As is described in Chapter Five, this issue was fundamental to the design of the Stavanger Wargame.

Second, between Europe and the United States, the Alliance's amphibious forces collectively offer an impressive capacity with the potential to deter aggression. Figure 2.1 shows these forces. However, while quantity may be sufficient, quality varies greatly. In the near term, participants stressed that NATO's amphibious forces need to emphasize ship and landing force readiness and the ability to aggregate into scalable multinational formations that "plug into" NATO's command structure. This point reinforced the need to continue wargaming to find a candidate C2 solution that integrates forces from ALES nations.

Finally, the workshop identified an opportunity to enhance the role of amphibious forces within NATO by highlighting the issue ahead of the 2018 Brussels Summit. In the months after the workshop, NATO officials developed the Allied Maritime Posture initiative with objectives for allied amphibious forces, which was approved in time for the defense ministerial meeting prior to the summit and reinforced in the Brussels Summit Declaration approved by heads of state and government:

> We are reinforcing our maritime posture and have taken concrete steps to improve our overall maritime situational awareness. We have prepared strategic assessments on the Baltic and Black Seas, the North Atlantic, and the Mediterranean. Through

---

[13] The event took place at RAND's Arlington, Virginia, facility and included leaders and staff from NATO organizations, the Office of the Secretary of Defense, the U.S. Department of State, the U.S. Marine Corps, the U.S. Navy, allied militaries, and think tanks.

Figure 2.1
Amphibious Forces in Europe

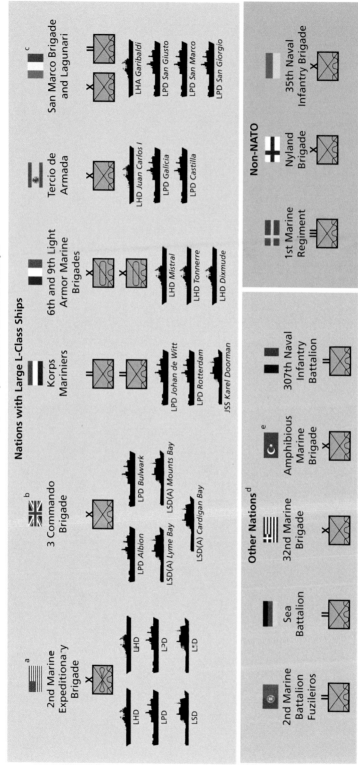

NATO and Partner Amphibious Forces in Europe

a The 2nd MEB, which could deploy with between six and fifteen ships, prepares for amphibious operations in Europe. Other Marine Corps forces may also be employed in Europe based on mission needs.
b One of two Queen Elizabeth class carriers, not pictured, will be capable of performing an LPH role in the 2020s.
c Italy maintains two amphibious elements. In addition to the navy's marine brigade, the army maintains an amphibious battalion also designed to operate from navy ships. The aircraft carrier Cavour, not pictured, could serve as a command ship for amphibious operations while carrying a contingent of marines
d These nations currently lack large L-class ships. Their landing forces are employed through a variety of means such as surface combatants, tank landing ships, landing craft, specialized boats, and amphibious ships belonging to other nations.
e Turkey is procuring an LHD in the 2020s.

an enhanced exercise programme, we will reinvigorate our collective maritime warfighting skills in key areas, including anti-submarine warfare, **amphibious operations**, and protection of sea lines of communications. The posture will also ensure support to reinforcement by and from the sea, including the transatlantic dimension with the North Atlantic being a line of communication for strategic reinforcement.[14]

It is in this context that the third and final ALES event regarding C2, the Stavanger Wargame, took place, as described in Chapter Five. The remainder of this report describes the two wargames and seminar and concludes by reflecting on NATO's progress and prospects for amphibious integration and interoperability.

---

[14] NATO, 2018; emphasis added.

CHAPTER THREE

The Naples Tabletop

Purpose

On June 20–21, 2017, the Naples Tabletop examined current and potential C2 constructs for multinational maritime and amphibious forces in support of NATO. RAND researchers designed and facilitated this TTX to evaluate existing naval C2 constructs and begin the process of developing improved concepts to be tested and refined during future conferences, wargames, and live or simulated exercises with actual headquarters and forces. During this TTX, participants explored the C2 challenges and dynamics that could emerge in a conventional conflict against a near-peer adversary that included a large-scale amphibious operation in the North Atlantic region.

This role-playing exercise was designed at the JFC, CFMCC, and ATF levels of decisionmaking, with game moves centered on C2 and directly related issues. The scenario, detailed in Appendix B, did not test the overall viability of amphibious operations in the European theater. Additionally, the game remained at the strategic and operational levels of warfighting. TTX moves and structured discussions excluded tactical and system-specific considerations unless they were directly related to C2 constructs and concepts.

The commanders (and each of their deputies) of JFC Naples, SFN, and MARCOM participated alongside national contingents and senior representatives from NATO's International Military Staff and Allied Command Transformation (ACT). The group's expertise and seniority enabled a realistic discussion on how the Alliance could arrange C2 for the maritime and amphibious aspects of an MJO+.

Event Design and Execution

Event Design
Before the exercise, a Move Zero (which is discussed further below) served as an abbreviated force generation process and a venue for discussing national caveats. The first

game move was designed to examine a C2 construct whereby two maritime commands—a JFC (with MARCOM as its CFMCC) and SFN—divided responsibilities across a joint operations area (JOA).[1] A second move provided participants the opportunity to design and explore an alternative, unified construct for C2 of maritime forces.

The scenario was set in late summer 2018 and postulated a Red invasion of the Baltics that escalated to include a Red seizure of three airbases along the northwestern coast of Norway. The full scenario is provided in Appendix B.

Participants were tasked to develop a concept of operations (CONOPS) with a supporting maritime and amphibious C2 construct to establish maritime control within the JOA, degrade Red A2/AD capabilities, and conduct amphibious operations to retake the three captured airfields. The TTX design focused on maritime and amphibious operations, stopping short of exploring transition ashore and follow-on land operations.

Move Zero

Before actual TTX play, each national contingent was provided with a comprehensive joint statement of requirements (CJSOR) and asked to specify which of their naval forces would be available for operations in the North Atlantic and which would be reserved for other theater needs or as a national reserve within the context of the scenario.[2] Table 3.1 summarizes the ships, submarines, maritime patrol aircraft, and landing forces designated as operationally available by game controllers.[3] According to interviews with NATO officials, a force of this size would be expected for an MJO+. The game design made two major assumptions to arrive at the CJSOR:

1. The United States would provide three carrier strike groups (CSGs), shipping for a Marine expeditionary brigade (MEB) afloat, and an additional Marine expeditionary unit or independent amphibious element.
2. Remaining TTX nations would provide between one-half and three-quarters of their navies for sourcing.[4]

In addition to sourcing their forces, participants were also asked to identify national caveats based on political, capability, or doctrinal considerations.[5]

---

[1] Since NATO does not currently have fully evolved plans for an MJO+ level of conflict, there is no official, agreed-upon, and documented construct, although interviews before the game indicated that the divided JFC-SFN design would be a likely starting point for operational planners should a large-scale conflict take place in the immediate future.

[2] A CJSOR is a list of approved forces and capabilities required for a particular operation.

[3] Sources used to derive the CJSOR included International Institute for Strategic Studies, *Military Balance 2017*, online database, 2017; IHS Markit, *Jane's*, online database, undated; and discussions with MARFOREUR/AF staff in May 2017 in Brussels, Belgium.

[4] This aggressive assumption made a large force available for the examination of C2 in a large-scale operation.

[5] Instructions sent before the TTX and repeated at the event clarified that national caveats would not represent official positions and were to be used solely for game play.

Table 3.1
Naples Tabletop Summary of Available Assets

Asset	Total Operationally Available for TTX	ESP	FRA	GBR	ITA	NLD	NOR	PRT	USA	Other NATO
Aircraft carriers	5	—	1	—	1	—	—	—	3	—
Amphibious ships	27	2	2	4	2	2	—	—	15	—
Major surface combatants	68	6	10	10	9	3	3	3	17	7
Mine warfare ships	40	3	6	8	5	3	4	—	2	9
Attack submarines	29	2	3	4	4	2	3	1	8	2
Maritime patrol aircraft	52	—	3	—	—	13	3	—	24	9
Battalion landing teams	9.5	1	1	2	1	1	—	0.5	3	—

Games One and Two

On the first day, two parallel games were planned to test the current "as is" C2 construct depicted in Figure 3.1. Players in each game were further subdivided into teams representing JFC and SFN. The participants acted as the leadership of the JFC and SFN, and they did not play their own national and organizational roles.[6] Each game was designed to be played through two moves of the same scenario. The first move would focus on the maritime stage of the operation, achieving sea control and protecting the SLOCs, while the second addressed the execution of large-scale amphibious operations. The game's participants were tasked to develop a CONOPS with four elements:

1. support and C2 relationships between JFC and SFN
2. assignment of tasks and allocation of forces
3. designation of the CATF and CLF
4. adjustments to battlespace geometry.

---

[6] Games 1 and 2 were designed as identical experiments. Individuals were assigned seats to ensure balance in expertise, nationalities, and ranks. A White Cell common to both games answered requests for information and received outbriefs alongside RAND game controllers. This White Cell played the role of SACEUR and the combined force air component command (CFACC), combined force land component command (CFLCC), and combined force special operations component command (CFSOCC).

Figure 3.1
Baseline Naples Tabletop Command and Control Construct

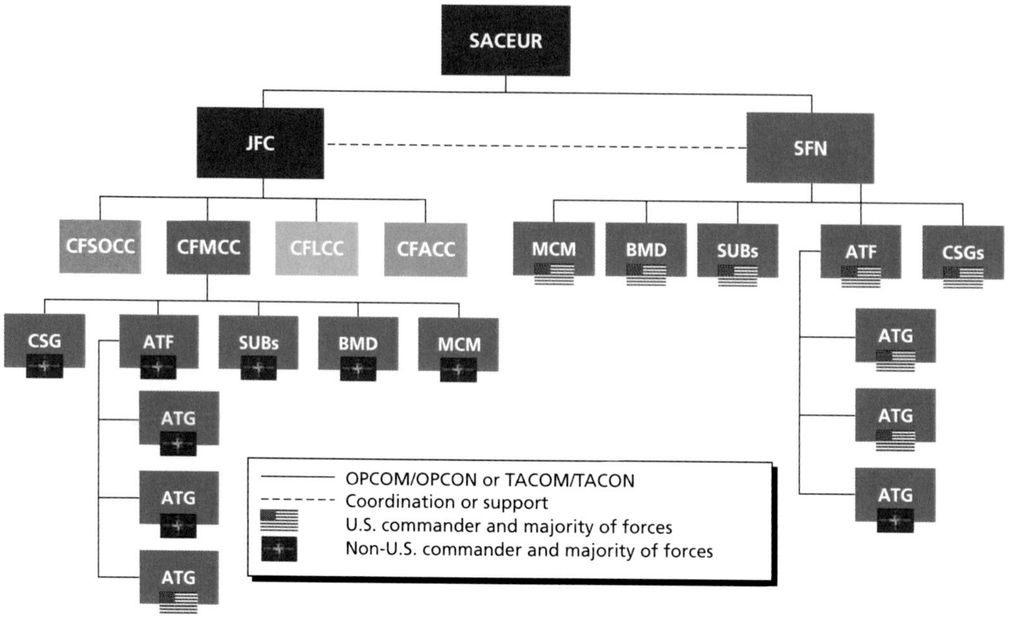

NOTES: OPCOM/N = operational command/control; TACOM/N = tactical command/control.

The original design for the second day of the TTX was for the players to collectively evaluate the performance of the "as is" C2 construct employed on the first day and either adjust it or devise a new one. The scenario would then be replayed by a single team, with players assuming roles that most closely matched their real-world national or NATO responsibilities. The alternative construct in Figure 3.2 was designed as a potential starting point for the second day. The intent was to test a divided JFC-SFN construct on day one, with participants incorporating supported-supporting relationships into their CONOPS. The second day would test a unified construct under a single CFMCC.

The next section discusses participant deliberations during this two-day event and documents the arrived-at CONOPS and associated C2 design from each of the two games.

Event Execution

Move Zero took place during the first morning, with national contingents sourcing their forces to the North Atlantic operation and identifying national caveats to their employment by SACEUR. Discussion on the first day produced vigorous debate over the characteristics of potential C2 constructs. Teams in both parallel games (each consisting of a JFC-SFN combined planning group) quickly and adamantly rejected the

Figure 3.2
Alternative Naples Tabletop Command and Control Construct

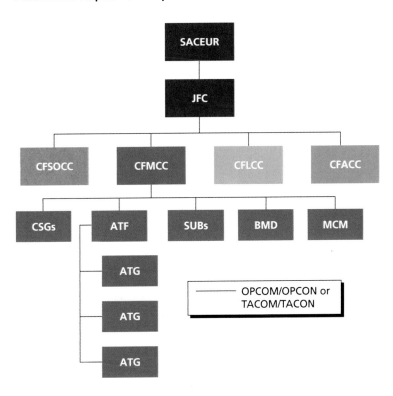

"as is" design in favor of a unified command organization. Thus, the teams managed to address only the initial maritime phase of the campaign. The RAND control team therefore decided to use most of the second day to work through the amphibious operations phase. To conclude the second day, RAND facilitated a structured discussion on C2 issues that emerged during the TTX among all game participants.

Move Zero

After determining that most of their forces would be made available for NATO's maritime campaign in the North Atlantic, national contingents conducted an abbreviated force generation process by sourcing their forces to JFC and/or SFN. Table 3.2 presents their decisions.

The national contingents took four different approaches to sourcing the operation:

1. Allocate forces to SFN only.
2. Allocate forces to JFC only.
3. Allocate forces across JFC and SFN.
4. Provide forces for SACEUR's allocation.

Table 3.2
Naples Tabletop Move Zero Decisions on Sourcing

STRIKFORNATO (North Atlantic Operation)										
Asset	Total	ESP	FRA	GBR	ITA	NLD	NOR	PRT	USA[a]	Other NATO
Aircraft carriers	3		—	—	—			—	3	
Amphibious ships	15		—	—	—			—	15	
Major surface combatants	24		2	4	1			2	15	
Mine warfare ships	5		3	—	—			—	2	
Attack submarines	12		—	2	1			1	8	
Maritime patrol aircraft	20		3	—	—			—	17	
Battalion landing teams	3.5		—	—	—			0.5	3	

Joint Force Command (North Atlantic Operation)										
Asset	Total	ESP	FRA	GBR	ITA	NLD	NOR	PRT	USA	Other NATO
Aircraft carriers	2	—	1[b]	—	1	—	—		—	—
Amphibious ships	12	2	2	4	2	2	—		—	—
Major surface combatants	24	1	6	4	3	3	3			4
Mine warfare ships	19	1	3	4	3	—	4			4
Attack submarines	7	—	2	—	1	2	—			2
Maritime patrol aircraft	11	—	3	—	—	—	3			5
Battalion landing teams	6	1	1	2	1	1	—			—

[a] USA provides 2nd Marine Air Wing
[b] French TACOM

In the last case, game controllers assigned a nation's forces either entirely to SFN or entirely to the JFC.[7] Three trends manifested during force generation discussions:

1. Aircraft carriers, amphibious ships, and major surface combatants were divided almost evenly between JFC and SFN, although SFN's forces included U.S. capabilities (the CSGs and amphibious ready groups) that participants acknowledged provide more protection and lethality.
2. The U.S. capacity in submarines and maritime patrol aircraft meant a greater SFN capacity to execute subsurface tasks.
3. European mine countermeasures (MCM) capacity exceeded that of the United States and was largely placed under the JFC.

Next, as shown in Figure 3.3, national contingents identified caveats to the employment of their forces by SACEUR. These included a range of political and doctrinal considerations. Most could be readily accommodated, but a few could have affected CONOPS considered during the TTX. Several nations retained control of their strategic forces and national cybercapabilities. Other countries limited the use of specific platforms and capabilities (e.g., cruise missiles). One nation placed a time limit on its transfer of authority, while another did not have any national caveats.

Reflecting how crisis deliberations are often conducted in a coalition or alliance setting, the TTX dictated that political and military decisions be made with limited information. Given a chance to revisit their sourcing decisions and national caveats at the end of the game, some participants stated that they would have approached Move Zero differently had they known that, during each game's initial discussions, SFN would be assigned the CFMCC role, with MARCOM largely serving as an adviser or assumed to be playing a limited supporting role.

At the end of Move Zero, participants were divided into two parallel games and a White Cell common to both games.

Game One

Upon examining the baseline C2 design (the JFC-SFN divided construct) provided by game controllers as part of the scenario, players quickly decided that it would not work. Discussion focused on how this divided C2 construct violates the military principle of unity of command. For example, shifting assets to accommodate operational priorities at each stage of a campaign (i.e., as the operation's main effort shifted from maritime to amphibious tasks) would require either a burdensome JFC-SFN negotiation or SACEUR decision on matters well below the strategic level. Players also stated that this bifurcated arrangement would require an extensive effort to deconflict across geographic and functional boundaries to avoid exposing gaps and seams.

---

[7]   Game controllers also played the role of Spain, which was unable to participate because of concurrent national commitments.

Figure 3.3
Naples Tabletop Move Zero Decisions on National Caveats

	National Caveats	Notional for Game Purposes Only

- **France**
  - SSN (nuclear-powered submarine) is under national submarine operating authority
  - OPCOM of French forces remains French
  - TACOM of carrier remains French
  - National control of cruise missile strikes
- **Italy**
  - One frigate has to stay attached to MCM ships for escort and support
  - Battalion landing team (BLT) must have its own area of responsibility and geography
- **Netherlands**
  - No caveats
- **Norway**
  - Special forces stay under national command
  - Maritime patrol aircraft, submarines, and the joint headquarters (HQ) stay under national command unless specifically requested by NATO
- **Portugal**
  - Marines for raids or small-scale operations only (additional company of marines available for operations aboard allied LPD)
  - Two frigates and two patrol vessels to remain together (both to transport marines)
  - Forces' transfer of authority valid for six months
- **United Kingdom**
  - Provides all forces to NATO but reserves the right to withdraw for homeland defense emergency (if bastion threatened)
  - Retains control of all strategic weapons systems
  - Retains control of all cyberweapon system employment
- **United States**
  - SSNs may be removed to protect homeland

Game One participants therefore engaged the White Cell to request the establishment of a single CFMCC, although there was no consensus on which command should serve in this role. Discussions regarding candidate CFMCCs identified three organizations. Each would face limitations if tasked to serve as the CFMCC in a campaign of the size and complexity required by this scenario:

1. **SFN** is a NATO command specializing in integrating U.S. Navy and U.S. Marine Corps forces into NATO operations. Participants agreed that SFN lacks the full capacity and expertise needed to serve as CFMCC in a campaign of this size and complexity. In addition, for some participants, placing SFN in the CFMCC role raised political concerns stemming from their perception of how SFN is not fully integrated into the NATO Command Structure.

2. **MARCOM** is fully integrated into the NATO Command Structure and is tasked with acting as the maritime component in a major joint operation. How-

ever, participants recognized that, despite this tasking, MARCOM has limited capabilities for commanding large, high-end maritime forces during wartime.

3. **HRF(M)** headquarters is an option that was briefly raised, but none of the current HRF(M) designees possess the full capability required for an MJO+. Some of the players had a limited awareness of the HRF(M) headquarters construct.

The White Cell accepted the recommendation for a single CFMCC and directed that SFN, which most participants acknowledged has the most capacity of the three options, lead the entirety of the operation. The group proceeded with an SFN-led construct despite concerns over span of control, political implications, and the realization that SFN would likely require augmentation. Players then developed their CONOPS and associated C2 below the CFMCC level. The arrived-at CONOPS involved a three-phase operation:

1. Secure SLOCs.
2. Conduct joint targeting and shaping of the amphibious objective area (AOA).
3. Conduct amphibious assault.

To accomplish these phases, players constructed a theater ASW task force, five CSG task forces, and an ATF. Each of the five CSG task forces (three from the United States and one each from France and Italy) fell directly under SFN, although France retained tactical command (TACOM) per national caveats specified in Move Zero.

In designing their detailed CONOPS, Game One participants debated the designation of the main effort in each phase, the appropriate pace of operations, and conditions for phasing. Most participants emphasized the importance of degrading Red submarines and the A2/AD threat before transitioning to Phase 3. One of the players expressed this point of view by stating, "Until the coastal defense cruise missiles [CDCMs] and surface-to-air missiles are gone, we are not going to move any amphib near [the threat envelope of these systems]." The group estimated that it would require 90 days to accomplish this degree of Red attrition before it could initiate the amphibious assault. A competing viewpoint pressed for a faster tempo: "The longer Red has to mine the harbors, the more difficult it will be later [to conduct the assault]. Red's ongoing reinforcement of current positions would also make a delayed landing more difficult."

Players requested CFACC and CFSOCC assets to support their operations in Phase 2, but the White Cell and game controllers responded that most of these forces were committed to the main effort in the Baltics.[8] For this reason and others (e.g., Norway's inherent familiarity with the terrain, and political considerations), the group

---

[8] Players proceeded to rely on CSG air power and a U.S. Marine air wing positioned in southern Norway. Special operation forces were a recognized shortfall.

deemed it essential to ensure early integration with Norway's three-star joint headquarters to support advance force operations and maintain situational awareness.

For the amphibious assault in Phase 3, the group developed a plan intended to transition smoothly from the larger maritime operation:

1. The CATF and CLF were colocated and would most likely be a three-star U.S. admiral and three-star U.S. general reporting to the CFMCC. There were differing views on the CATF/CLF's location (e.g., large L-class amphibious assault ship, purpose-built command ship, or ashore), but all agreed on the need to plan and communicate across several multinational ATGs. Players expressed concern over U.S. ability and willingness to conduct the operation on a NATO SECRET network, among other CIS interoperability issues that could disrupt operations.

2. The ATF was made up of one or two ATGs. Planning focused on generating forces ashore as quickly as possible. While there was some discussion on integrating other nations' amphibious forces early in the operation, players felt that a UK-U.S. ATG serving as the lead element would most rapidly develop combat power. Other national amphibious forces would serve as follow-on forces. Units would shift missions as objectives were achieved.

3. Participants stressed the importance of conducting a full rehearsal given the Alliance's lack of recent experience in a maritime campaign of this scope and complexity.

At the end of the discussion, participants reemphasized that a single maritime commander is needed because the SLOC protection, shaping, and amphibious assault mission sets are inextricably linked. A single maritime commander could more efficiently distribute and prioritize forces across mission areas based on operational phasing.

Game Two

As was the case in Game One, participants in Game Two quickly and adamantly rejected the baseline C2 design provided at the onset of planning. Players did not understand why maritime C2 would be divided between two commands and initially agreed to unified C2 of maritime forces under SFN.

Game Two also resulted in a three-phase operational design:

1. Establish broad-area sea control and begin shaping the AOA/JOA.
2. Establish sea and air control of the AOA and complete shaping.
3. Execute amphibious assault.

The focus of operations in Phases 1 and 2 was on ASW, sea control, air defense, and shaping Red A2/AD capabilities. As shown in Figure 3.4, Game Two participants, like their counterparts in Game One, established a theater ASW (and undersea war-

Figure 3.4
Naples Tabletop Game Two Command Structure

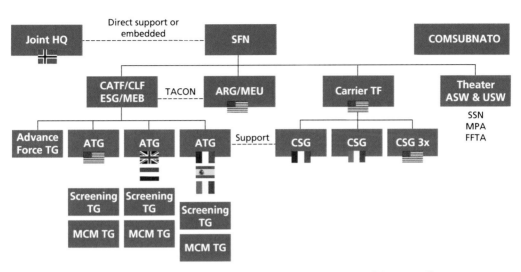

NOTES: COMSUBNATO = Commander, NATO Submarine Forces; ESG = expeditionary strike group;
FFTA = frigate, towed array; MPA = maritime patrol aircraft.

fare, or USW) task force and a CATF/CLF reporting to SFN. The C2 design differed across the two games in that the participants of Game Two established a consolidated carrier task force commander as an intermediate command between SFN and the CSGs to direct the air and missile defense and strike functions. Game Two participants also held discussions about providing SFN with an amphibious ready group/ Marine expeditionary unit (ARG/MEU) to act as an independent reserve force.

C2 of ASW was the critical focus for planning the maritime battle. Participants agreed that standing NATO and multinational ASW constructs work well in peacetime (e.g., sea space management by the COMSUBNATO), but they were unsure whether these mechanisms would scale up to perform effectively in an MJO+ (e.g., theater-wide prioritization of ASW forces).

As was the case in Game One, participants emphasized the role of Norwegian forces and resultant C2 issues, something the game design did not properly capture. Norwegian forces offer significant capability in this scenario, including the three-star joint force headquarters. Maritime forces would need to coordinate air defense, coastal defense, and special operations with Norwegian forces during Phases 2 and 3 of the campaign, along with ground operations after the Phase 3 landings.

For the amphibious operation, participants arrived at the following plan:

1.   The C2 design allocated battlespace to national elements or bilateral/trilateral elements with habitual relationships. The aim was to provide unity of com-

mand for each of the three amphibious objectives while leveraging habitual relationships among amphibious elements (the United Kingdom and the Netherlands; Spain and Italy).

2. Participants agreed that the United States likely had the only capability to serve as CATF/CLF and concurred in that C2 construct as long as national elements retained C2 of their subordinate forces, maintained unit integrity, and were assigned distinct operating areas. The U.S. ESG and Marine expeditionary force commanders were mentioned as the most likely candidates to assume the CATF/CLF role.

As the game progressed, it became clear that the scope and scale of maritime C2 required by this scenario were beyond SFN's capacity, notwithstanding the fact that this command has the most capacity of the candidates discussed. SFN needed to "offload" tasks, such as controlling SLOCs and open-ocean ASW, but there was no existing maritime command to hand off those tasks to. At game's end, participants still endorsed the concept of a single maritime command exercising overall C2, but they felt that some division of labor would be required to deal with the multiplicity of tasks across the large operating area depicted in Figure 3.5.

To summarize, players in both games swiftly rejected the baseline C2 design with command divided between JFC and SFN. Participants agreed to a unified command

Figure 3.5
Naples Tabletop Operating Area

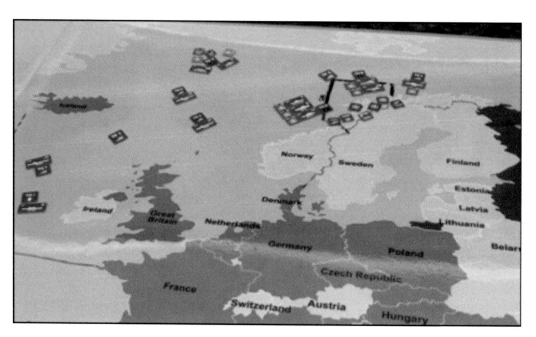

with SFN acting as CFMCC but acknowledged political and practical constraints on SFN acting in this role during an actual operation—at least as SFN is currently configured. Both games developed similar three-phase CONOPS: establishment of SLOC security, shaping of the AOA, and execution of the amphibious assaults. The amphibious C2 design allocated tasks to national or bilateral and multinational forces with habitual relationships.

Participant Observations and Recommendations

Caution must be shown in drawing conclusions from a single TTX. However, the exercise yielded observations for consideration by senior leaders, further analysis during allied planning, and examination in future wargames. Chapter Six documents progress on some of these issues since the event.

Observation 1: *Participants assessed the current C2 construct for large-scale allied maritime operations as unsuitable for an MJO+ against a near-peer adversary.* There was no consensus on the exact shape that the C2 construct should take, but all participants concluded that the existing construct was not appropriate for an MJO+. The downsides of a divided JFC-SFN command design noted by participants included

- the absence of a maritime commander below SACEUR with the authority to arbitrate competing force allocation demands across maritime and amphibious missions
- the effort needed to deconflict across geographic and functional boundaries to avoid exposing gaps and seams.

While participants quickly agreed that unity of command was a key factor in the design of a C2 construct, they also acknowledged that the complexity of maritime operations in an MJO+ may exceed the span of control of a single maritime command.

Observation 2: *Neither SFN nor MARCOM is currently postured to serve as the Alliance's CFMCC for an MJO+ such as this TTX's North Atlantic scenario, although SFN currently has the most capacity to manage large-scale allied maritime operations.* There are existing national, bilateral, and multinational organizations capable of fighting aspects of the maritime and amphibious battle, but none can exercise overall C2 of the entire maritime campaign. While TTX participants explored the potential of SFN overseeing operations in the North Atlantic, they recognized that this headquarters is not trained, equipped, organized, or funded beyond its primary mission of integrating U.S. naval and amphibious forces into allied operations. SFN (as configured today) would struggle to manage the range of complex functions—from open-ocean ASW to coastal defense—over an operating area covering the maritime approaches and environs of the northern European theater. MARCOM would be even more stressed if tasked beyond its current peacetime responsibilities and limited wartime capacity.

Observation 3: *NATO has the capability to execute C2 of amphibious forces at the national or bilateral ATG or ARG/MEU level, but it would face challenges in controlling larger-scale operations.* Maritime and amphibious forces maintain habitual relationships and are comfortable working within a larger NATO or U.S.-led organization as long as national forces are employed as integral entities. However, exercising C2 of larger formations and coordination across ATGs requires organizations and capabilities that either do not exist or do not maintain the requisite capacity and expertise.

Observation 4: *As currently postured, there is no immediately ready source of a CATF overseeing several multinational ATGs.* The U.S. ESG appears to be the best potential candidate to fill this role, but several steps are needed to prepare for an operation on the scale of an MJO+. Potential focus areas include enhancing communications interoperability, conducting MJO+ exercises, establishing an allied requirement for a standing multinational CATF/CLF staff, and identifying suitable flagships to host the ATF and landing force headquarters.

Observation 5: *Experience in the C2 of large-scale maritime and amphibious operations against a near-peer adversary has atrophied across NATO, including within the U.S. Armed Forces.* Improved capabilities are needed across the DOT-MLPF-I spectrum. Participants agreed on the need to rebuild expertise through education, training, and exercises.

Observation 6: *European allies have made investments in support of maritime C2 capabilities that some TTX participants found to be underutilized or disregarded by NATO and U.S. planners.* These include ship modifications to support maritime command staffs, communication suite upgrades, and the Alliance's designated HRF(M) headquarters. Some TTX participants from allied nations sought clarification on whether these capabilities should be maintained, postured differently, or stood down based on their value to potential NATO contingencies such as the examined scenario.

Observation 7: *National caveats included a range of political and doctrinal considerations. Most could be readily accommodated, but a few could have affected CONOPSs considered during the TTX.* Several nations retained control of their strategic forces and national cybercapabilities. Another set of countries placed limitations on the use of specific platforms and capabilities (e.g., cruise missiles). One nation stated that its landing force must have its own area of responsibility, while another put a six-month time limit to the transfer of authority.

Participants made several proposals to address issues raised during the TTX. While there was insufficient time within the two days to explore these proposals in depth or gain unanimous consent, the following ideas had support from at least a portion of the participants and should be explored further.

Recommendation 1: *Develop a new concept for maritime C2 within the NATO Command Structure.* Participants from ACT and Supreme Headquarters, Allied Powers Europe pointed to ongoing efforts to reexamine the NATO Command Structure as a venue to gain support for such an effort. Regardless of NATO's higher-level command structure reforms, however, issues on how to organize within a CFMCC (and ATF) will need continued attention.

Recommendation 2: *Rigorously test the ability of existing C2 organizations to conduct maritime and amphibious C2 in MJO+ operations in future wargames and exercises.* While participants generally felt that existing maritime and amphibious C2 organizations were inadequate to the task, they also acknowledged that their capabilities had not been fully tested. Some participants felt that enhancing existing organizations would be a faster and more efficient way to develop the capability than trying to create new commands.

Recommendation 3: *Develop a standing, multinational two-star headquarters to serve as CATF/CLF for major amphibious operations.* Some participants felt this was the best way to address the deficiency in structure and expertise in this area. They did acknowledge the difficulty in resourcing any new headquarters, particularly absent a formal allied requirement.

Recommendation 4: *Reduce the number of exercises and refocus exercise objectives on improving C2 of maritime and amphibious forces at the operational level.* Flag officer participation is crucial to advancing understanding and capabilities to conduct C2 at this level.

Recommendation 5: *Provide more frequent and challenging training for maritime and amphibious headquarters that would exercise C2 in an MJO+.* That training should include broader and deeper scenarios against a realistic high-end threat.

Chapter Summary

This TTX explored the C2 challenges and dynamics that could emerge in a conventional conflict against a near-peer adversary that included a large-scale amphibious operation in the North Atlantic region. TTX participants—today's maritime leaders within the Alliance—came away with a clearer understanding of the requirements and challenges they face in evolving the current maritime and amphibious C2 constructs to meet the needs of the emerging security environment.

For U.S. participants, the TTX reinforced the importance of including partner nations in plans, operations, and exercises. As the TTX progressed, the value of partner capabilities became apparent, as did the challenges of achieving synergy and integration within a complex multinational C2 construct.

This 2017 TTX began a much-needed discussion about what NATO's maritime C2 construct should be in an MJO+ against a near-peer competitor and identified several potentially important insights. More broadly, the key takeaway from the Naples Tabletop was that as NATO reevaluates its overall command structure in response to the emerging threat environment, it should revamp its maritime C2 construct to address requirements for fighting a high-end conflict. In the months following this TTX, the Alliance developed a plan to create JFC Norfolk to oversee potential high-end, maritime-focused operations in the North Atlantic; the proposal was formally adopted at the 2018 Brussels Summit.[9]

The Naples Tabletop did not, however, explore in detail how a capable and competent Red could undermine the C2 constructs that were explored, nor did it consider alternative amphibious constructs that could "plug into" NATO's evolving command structure. ALES stakeholders therefore committed to reconvening, first at the Northwood High North Seminar to discuss amphibious C2 options, and next at the Stavanger Wargame to explore a candidate amphibious C2 construct prior to injection into exercises.

---

[9]   JFC Norfolk is designed to operate in close coordination with the reconstituted U.S. 2nd Fleet. See U.S. Navy Office of Information, "CNO Announces Establishment of U.S. 2nd Fleet," NNS180504-15, Washington, D.C.: U.S. Navy, May 4, 2018.

CHAPTER FOUR

The Northwood High North Seminar

Purpose

The next ALES event was held in Northwood, United Kingdom, on November 21–22, 2017. This seminar was cohosted by the Commandant General, UK Royal Marines, and Commander, MARFOREUR/AF, and featured two related topics. To begin, participants discussed recent developments and challenges in the High North security environment. The workshop then transitioned to further exploration of C2 at the amphibious forces level. Continuing with ALES's focus on MJO+ that began at the Naples Tabletop, the workshop focused on the organization and employment of an allied ATF, with subordinate national and bilateral ATGs, for major combat operations. These deliberations occurred against the now familiar North Atlantic crisis scenario developed by RAND and documented in Appendix B.

Participants included one- and two-star flag and general officers from ALES nations, now joined by their Norwegian colleagues. NATO International Military Staff, ACT, MARCOM, and SFN also sent senior officers or their representatives. This level of participation facilitated a politically informed exchange of views about the High North and provided the expertise needed to identify a candidate amphibious C2 construct for further exploration at the Stavanger Wargame that would take place in June 2018.[1]

Event Design and Execution

Event Design
The agenda for the first day began with briefings on the High North security environment and the status of NATO's graduated response plans. External speakers from think tanks made unclassified presentations while military officers from High North nations and MARCOM presented classified information regarding national and Alliance adaptations to recent security challenges.

---

[1] One of the themes from the Naples Tabletop and prior ALES events was amphibious leader recognition of the need to meet on a reoccurring basis at the flag and general officer level.

RAND set the scene for amphibious C2 deliberations by briefing the outcomes of the Naples Tabletop and facilitating presentations by experts on maritime and amphibious doctrine. Representatives from the Combined Joint Operations from the Sea Center of Excellence (CJOS COE) and the U.S. Naval War College provided briefings on amphibious and maritime doctrine and best practices, respectively.

On the second day, participants were divided into two groups for exploration of amphibious C2 constructs. With an up-to-date understanding of NATO and U.S. doctrine presented earlier, each group was asked to consider an alternative amphibious C2 construct. Group One examined what was later termed a centralized ATF concept; Group Two deliberated the advantages and disadvantages of a decentralized version. To conclude the seminar, the two groups reconvened to identify the appropriate construct for the Stavanger Wargame, based on a collective vision of what construct(s) should be further evolved and tested.

Event Execution
Doctrine Scene Setter
After discussions about the High North and a review of the Naples Tabletop, CJOS COE and U.S. Naval War College presenters emphasized the following principles as a prelude to deliberations regarding amphibious C2 concepts:

1. Maritime C2 follows the tenets of general C2 principles but is characterized by decentralized execution, use of adaptive and flexible organizational arrangement, and command by negation necessitated by the dispersed and multi-domain nature of the maritime fight.[2]
2. Maritime C2 must effectively nest within joint and combined C2 constructs; the CFMCC's participation on a JFC's operational planning teams, boards, bureaus, and cells is critical and requires appropriately experienced personnel.
3. Battlespace management is likely to cause friction in complex amphibious operations, particularly in relation to the air domain.
4. NATO and U.S. amphibious doctrine are generally aligned, although one aspect less explored in NATO doctrine is options for a common superior to the CATF/CLF, referred to as the commander, amphibious force (CAF) in U.S. doctrine.[3]

---

[2] Command by negation, referenced as command by veto in allied doctrine, is described as such:

In many aspects of maritime warfare, it is necessary to preplan the actions of a force to an assessed threat and to delegate some missions to a subordinate. Once delegated, the subordinate is to execute the mission without delay, always keeping the commander informed of the situation. The commander retains the power to veto any particular action. (NATO, *Allied Joint Doctrine for Maritime Operations*, AJP-3.1, Edition A, Version 1, Brussels: NATO Standardization Office, December 2016b).

[3] According to discussions at the seminar and in subsequent wargaming, the term *CAF* may be removed from U.S. doctrine in forthcoming revisions.

Figure 4.1
Notional Centralized and Decentralized Amphibious Task Forces

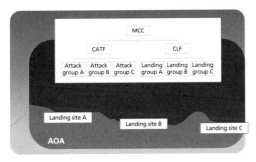 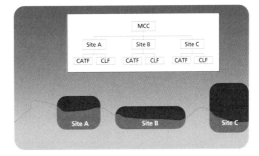

SOURCE: CJOS COE, slides from presentation in Northwood, United Kingdom, November 21, 2017.
NOTE: MCC = maritime component command.

One of the presentations included doctrinal representations of a notional centralized and decentralized ATF, depicted in Figure 4.1. After this brief, Group One would explore the centralized construct on the left-hand side, whereby a CFMCC would oversee one CATF and CLF pair, which would in turn lead operations across a single AOA. Group Two considered the construct on the right-hand side, with three AOAs for three separate CATFs and CLFs, which could report to the CFMCC or some form of intermediate command.

Group One's Discussion on the Centralized Amphibious Task Force

Group One considered the missions, functions, organization, and capability of an approximately two brigade–size multinational force tasked to complete an opposed landing against well-defended and potentially dispersed objectives under a single CATF and CLF. This construct posited the CATF and CLF as senior two- or three-star commanders sharing a combined amphibious force staff and embarked on a single platform. The CATF and CLF would coordinate but retain independent ability to request support from the CFMCC, depending on the phase of the operation. Key aspects regarding the envisioned use of this construct included the following:

1. CLF would most likely be a U.S. Marine Corps commander, but the CATF could be either a U.S. Navy or another NATO nation's flag officer, although no entity is currently organized to fulfill this role. Participants discussed options to address this deficiency, including a rotational CATF around a U.S. framework that could be augmented by multinational staff in time of crisis. Identifying a flagship with space to support the combined staff would be a priority in planning.

2. The centralized control model does not rule out constituting subordinate ATGs, and the link between national landing force elements and national force

amphibious shipping would be difficult to break. Transportation and loading practices generally reflect careful national planning consideration and disrupting this for anything other than an operational crisis would unsettle ship-to-shore movement and sustainment for landing forces.

3. A multinational and centralized fire support and airspace control authority could enable rapid access to fires and protection capabilities available across the force. This arrangement would require extensive planning, and thus a large multinational CATF/CLF staff and well-practiced tactical C2 procedures.

Overall, participants stated that the attractiveness of this model is the ability to delegate the entire AOA to a two- or three-star commander, which would free up the CFMCC to focus on other maritime functions and extensive coordination requirements with the JFC. Put another way, this paradigm centers on a CATF and CLF coming together to form a common and combined version of the CAF described in U.S. doctrine. This could engender unity of command and, depending on the situation, might sufficiently account for span of control. Potential limitations of the centralized ATF include the manning required for the multinational staff and the identification of a flagship with sufficient capacity to support the C2 and CIS functions.

Group Two's Discussion on the Decentralized Amphibious Task Force

Group Two considered a more decentralized task organization than Group One. Three separate CATF/CLF pairings were associated with three distinct objective sites, all answering to the CFMCC or some form of intermediate authority between the CFMCC and ATF. Using the Naples Tabletop scenario as a reference point, participants assumed that each CATF and CLF would likely be a one-star command (or at most a two-star command) with commensurate staff.

1. The presumption was that amphibious forces with habitual bilateral or single-nation training relationships would make up the CATF/CLF commands (i.e., ATGs) and subordinate naval and landing forces. Thus, this model comports with current force structure.[4]

2. Each CATF/CLF would likely be assigned its own AOA, and each respective AOA would be designed with differing characteristics based on the tactical situation. This construct seemed well suited to situations where objective sites were geographically separated by larger distances. However, participants noted that having multiple AOAs in close geographic proximity could make it dif-

---

[4]   Forces from the United Kingdom and the Netherlands maintain a habitual relationship and operate as an integrated United Kingdom Netherlands Amphibious Force. Spain and Italy have a similar relationship through SIAF/SILF construct. Spain may also embark Portuguese marines on Spanish ships, in which case the force becomes trilateral. France and the United States do not maintain integrated amphibious forces with other nations.

ficult to coordinate such functions as airspace management, fires, and missile defense, particularly against an adversary with high-speed, long-range weapons and platforms.

3. Participants noted C2 structural challenges related to span of control and management across the envisioned ATGs and generally agreed on what functions would need to be performed between the CFMCC and ATG echelons, such as
   a. determining sequencing and priority of objectives within the CONOPS, including designating forces as main or supporting efforts
   b. maintaining situational awareness of the entire amphibious fight (i.e., "seeing the big picture")
   c. reacting to enemy actions and adapting amphibious plans within the context of larger maritime operations
   d. managing maritime enablers and efficiently adjudicating requests for high-demand/low-density assets (e.g., maritime patrol aircraft) that would need to be shared across amphibious elements and AOAs. This would allow each CATF to focus on its priority of enabling the landing forces.

4. In a high-end scenario, the above functions could not be accomplished by a single one-star CATF/CLF selected from among equivalent adjacent units. Similarly, as had been explored in Naples, existing CFMCC options (e.g., SFN) lack the capacity and expertise to perform these functions from the top down. Given this realization, participants discussed four options for an organization or C2 construct that could plausibly perform these functions, each with strengths and weaknesses (see Table 4.1).[5]

Seminar participants recognized that selection of one of these options needs to be informed by the situation, including geography, enemy disposition, and the availability and state of allied forces at the time of a crisis. Despite this stipulation, a consensus emerged that some form of C2 construct needs to be selected to enable appropriate preparation (planning, manning, training, and equipping) for large-scale amphibious operations, with the recognition that any baseline solution will require situational adaptations. Any baseline C2 construct, therefore, should be designed to meet most likely operational conditions and structured to maximize flexibility.

---

[5]   The first option resembles the centralized construct considered by Group One.

Table 4.1
Alternative Amphibious Command and Control Structures

Options	Strengths	Weakness
Create an intermediate CAF-, CJTF-, or ESF-like command with command authorities.	Clearly identified by initiating directive and/or delegating directive to have command authorities. Better empowered to operate as OTC and quickly assign and reassign enablers as the situation requires.	Creates an additional level of command into the C2 structure, potentially slowing down pace of decisionmaking and force allocation. Requires significant personnel and material resources.
Create an intermediate CAF- or CJTF-like structure with coordination authorities but not command authorities.	Applies a CWC[6] approach while considering a more holistic understanding of CFMCC's objectives. Able to coordinate and facilitate subordinate CATFs/CLFs in response to requests and changing situation.	Without command authorities, subordinate CATFs/CLFs could be incentivized to go around that intermediate level and directly to the CFMCC to appeal for additional capabilities.
Augment CFMCC's staff with an amphibious operations cell.	Provides dedicated element focused on amphibious operations; amphibious staff function or cell would have full cognizance of CFMCC priorities and objectives.	Increased staff capacity may still be insufficient to manage complex, multi-ATG operations. Absence of command authority means CFMCC must still be involved in tactical and tactical-operational decisions concerning amphibious operations.
Enhance the CFMCC staff with amphibious-oriented personnel.	Simplicity: existing staff functions would remain unchanged; requires fewest personnel and resources.	Enhancing staff not likely to provide capacity required to fully coordinate major amphibious operations.

NOTES: CAF = commander, amphibious force; CJTF = combined joint task force; CWC = Composite Warfare Commander; ESF = expeditionary strike force; OTC = officer in tactical control.

Participant Observations

The following is a synthesis of the consensus observations regarding multinational amphibious C2, with the caveat that the discussion focused on a specific scenario and threat environment.

Observation 1: *The choice of ATF C2 arrangements should be dictated by the situation—mission, threat, battlespace geometry, and Blue force situation—rather than a predisposition toward either centralized or decentralized organization of the ATF.* Participants agreed that while a centralized construct offered more flexibility, the decentralized construct was more appropriate for missions with geographically dispersed objectives.

---

[6]   The CWC doctrine establishes a single officer in overall command of a naval task force or grouping, but delegates authorities for offensive and defensive functions to subordinate commanders who have the best capabilities to prosecute those functions. NATO, 2016b, p. 29.

Observation 2: *Amphibious C2 should exploit habitual relationships between and among national and bilateral amphibious forces to the maximum extent possible.* Links among national shipping, landing, and logistics elements (and, for some nations, air support) should be maintained to leverage the synergy and effectiveness of existing force structure.

Observation 3: *The centralized C2 construct provides flexibility and promotes interoperability.* However, there are limits to the degree of flexibility that could be achieved based on dependencies within extant forces (i.e., splitting national or habitually aligned bilateral elements might result in a net loss of overall effectiveness). Benefits of flexibility and interoperability can only be realized if there is a multinational amphibious staff able to effectively integrate the capabilities of the combined force. The staff must have the requisite manning, expertise, and training.

Observation 4: *Participants agreed that under either C2 construct, there is a need to have an appropriately staffed, trained, and equipped intermediate level of C2 between the national or bilateral amphibious elements and the CFMCC.* The consensus was that this level should include command authority along the lines of a CATF/CLF or the CAF concept currently in U.S. doctrine, although limiting this entity's authority to coordination (and not command) should be considered in planning. Participants envisioned the following characteristics for this structure:

1. The staff for this level of C2 would be multinational.
2. For the CLF role, the United States is best positioned to provide a two- or three-star officer and core staff from within II Marine Expeditionary Force.
3. For the CATF, there is no clear source of a corresponding navy officer and staff. NATO's HRF(M) rotational headquarters or the U.S. ESG are potential sources for this capability, but each would require augmentation and training.
4. Flagship availability and capacity would be an issue given that the size of staff for this command could be as high as 500 personnel.

Observation 5: *Critical questions concerning CFMCC-ATF-ATG relationships involve the prioritization, allocation, management, sequencing, and control of assets required for shaping, protection of the force, and integration of functions being executed at the maritime or joint force level (e.g., intelligence, surveillance, and reconnaissance; ASW; air defense; airspace control; and information operations).* Allocations and command relationships for these assets from the CFMCC to the ATF and its ATGs are situation dependent, but C2 constructs must provide the capability to plan, coordinate, and shift capabilities in a flexible and responsive manner.

Chapter Summary

Regarding the way ahead for the Stavanger Wargame, most participants favored testing the command structure identified as the centralized ATF, with a command element consisting of a two- or three-star-level CATF and CLF that would synchronize and integrate efforts and effects across the larger multinational ATF. The internal ATF organization could consist of individual naval and landing force elements reporting directly to this centralized CATF/CLF, subordinate ATGs each with their own CATF/CLF and AOAs, or some hybrid of the two.

CHAPTER FIVE

The Stavanger Wargame

Purpose

On June 20–21, 2018, general and flag officers from six allied nations and six NATO commands gathered in Stavanger, Norway, to explore a multinational amphibious C2 construct envisioned for employment against a near-peer adversary during an MJO+. The wargame was organized by MARFOREUR/AF and executed as part of ALES.

In 2016, ALES principals agreed that designing, wargaming, and testing C2 was imperative to generating a scalable multinational amphibious capability in support of NATO. During a series of events in 2017 culminating in the Northwood High North Seminar, ALES stakeholders developed a candidate C2 construct—the "centralized ATF"—for large-scale, multinational amphibious operations in an MJO+ conflict. As envisioned, the ATF consists of a colocated CATF and CLF, along with supporting multinational staffs and subordinate naval and landing force components. Figure 5.1, a simplified and representative C2 diagram, shows how this organization serves as an intermediate command between the CFMCC and smaller extant national and bilateral ATGs. The ATF directs, coordinates, and synchronizes the ATGs as a coherent force.

The purpose of this wargame was to explore the utility and function of the centralized ATF, better understand its viability in an MJO+, and examine the necessary actions NATO requires to plan, test, and implement the construct.

Event Design and Execution

Event Design

In consultation with MARFOREUR/AF and ALES stakeholders, RAND designed the wargame as a role-playing exercise set at the ATF and ATG levels of decision-making. Although the wargame's focus remained on ATF-to-ATG and ATG-to-ATG interaction, the RAND control cell also relied on a restricted-play Blue team to represent the CFMCC, which provided a broader theater and maritime context in which the centralized ATF operated and competed for resources.

Figure 5.1
Simplified Centralized Amphibious Task Force Construct

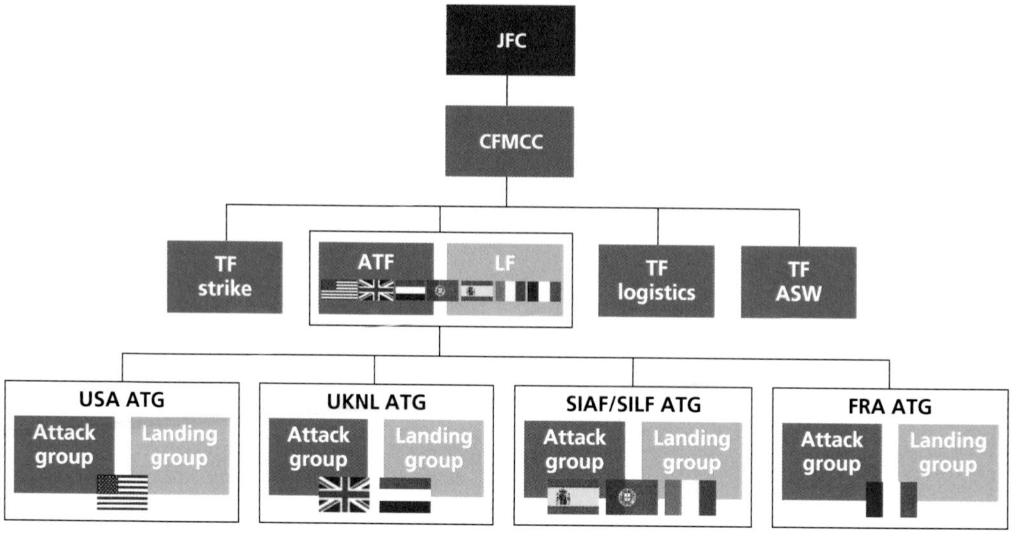

NOTE: This wargame centered on the ATF-to-ATG and ATG-to-ATG relationships. Higher-echelon maritime and joint C2 issues were the primary subject at the Naples Tabletop.

The scenario was set in mid-2019 and postulated a Red incursion onto the fictional island nation of Ostrov, a NATO member state located between Norway and Iceland. With Article 5 of the NATO Treaty invoked, SACEUR allocated most of the Alliance's maritime forces to the North Atlantic, including two dozen amphibious ships for the retaking of Red lodgments on Ostrov. The RAND control cell issued execute orders to the notional commands for the game, CFMCC-Northeast and ATF-Northeast, that directed the ATF to conduct three near-simultaneous assaults of objectives distributed along a coastline of several hundred miles. The scenario provided context to explore how a centralized, multinational ATF would apportion authorities, manage battlespace, allocate resources, and respond to enemy action in a complex, contested environment.[1] Appendix B provides the full scenario and the Red and Blue force lists.

The ATF team consisted of two groups: a CATF and staff led by a three-star U.S. Navy admiral, and a CLF and staff led by a three-star U.S. Marine Corps general. The two groups were colocated and each staff had multinational members. Four subordinate attack and landing groups, as shown in Figure 5.1, reported to the CATF and CLF, respectively.

---

[1]   Like the Naples Tabletop, this event did not simulate tactical warfighting nor the technical aspects of CIS performance or interoperability. These were, however, identified as areas requiring further planning during the discussion.

The wargame design consisted of three moves: prelanding (Move 1, or D-1), landing (Move 2, or D-Day), and postlanding (Move 3, or D+1). For each move, the ATF team was tasked to establish or refine its CONOPS and scheme of maneuver, allocate resources to each of the ATGs, and communicate any concerns to the CFMCC. The ATGs in turn developed more detailed plans to take assigned objectives, articulated national level capabilities and caveats, and made recommendations regarding force allocation, command relationships, control measures, and other C2-related elements of the ATF's plan. The Red team, played by RAND, took actions to pose C2-related challenges for the Blue team. Injects included Red actions that highlighted ambiguities in the ATF's overarching plan while stressing the ATGs to a point where support or reinforcement became necessary from forces belonging to higher and adjacent echelons.

This wargame's design differed from the Naples Tabletop in four respects:

1. As shown in Figure 5.2, the Stavanger scenario replaced Norway with the fictional island nation and NATO member Ostrov. Researchers created Ostrov and placed it between Iceland and Norway to focus participant discussion on

Figure 5.2
Stavanger Wargame Geography

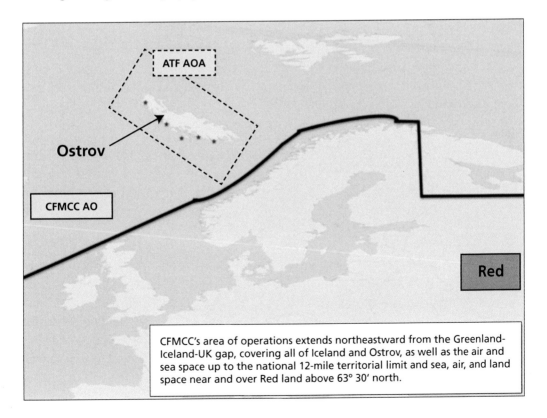

CFMCC's area of operations extends northeastward from the Greenland-Iceland-UK gap, covering all of Iceland and Ostrov, as well as the air and sea space up to the national 12-mile territorial limit and sea, air, and land space near and over Red land above 63° 30′ north.

amphibious assault, consolidation ashore, and rapid reconstitution at sea. This adjustment enabled the game to bypass issues related to integration with host nation forces and commands, which was explored to some degree during the Naples Tabletop.

2. In the Naples Tabletop, players were tasked with making plans for up to several weeks at a time. Moves in Stavanger, however, represented one-day increments. Researchers made this adjustment to shift focus from the full range of CFMCC issues to focus on ATF-ATG C2 integration and friction during an assault. This enabled injects that challenged the centralized ATF based on Blue's C2-related decisions and Red's actions.

3. The ATG echelon was introduced and consisted of four cells that had to interact frequently with one another and the ATF. Players were assigned based on nationality to correspond with the centralized ATF's envisioned national and bilateral subordinate force elements. Similarly, the CATF and CLF were played by the most likely real-life CATF/CLF combination for a high-end conflict.

4. A more detailed amphibious C2 construct and associated task organization was developed prior to the wargame at a planning meeting that included representatives from several ALES nations and NATO commands. Figure 5.3, expanding on the design articulated at Northwood, shows the C2 arrangements used for this wargame.

Event Execution

There was one notable difference between the wargame's design and its execution. The pace of game play permitted the compressing of Moves 2 and 3 into a single move, enabling a longer plenary session to explore DOTMLPF-I and political implications of the C2 construct explored during the game.

Move 1

After briefing the scenario, the facilitators also delivered guidance from SACEUR that the amphibious assault had to begin within one day and all objectives would need to be taken simultaneously, defined as landings commencing within 24 hours of one another. The combined CATF/CLF responded by designing an overarching CONOPS with the following components:

1. The initial landing on the afternoon of Day 1 would be conducted by the United Kingdom Netherlands Amphibious Force (UKNLAF) ATG against Andenes, Ostrov. The following morning, the Spanish Italian Amphibious Force/Spanish Italian Landing Force (SIAF/SILF) would land in Bodo and the U.S. ATG would assault Red forces in Tromsø. The matching of ATGs to objectives was based on each ATG's comparative advantages and relative combat power—the United States received the most difficult objective while differences in equipment informed UKNLAF and SIAF/SILF roles.

Figure 5.3
Stavanger Wargame Blue Command and Control Construct

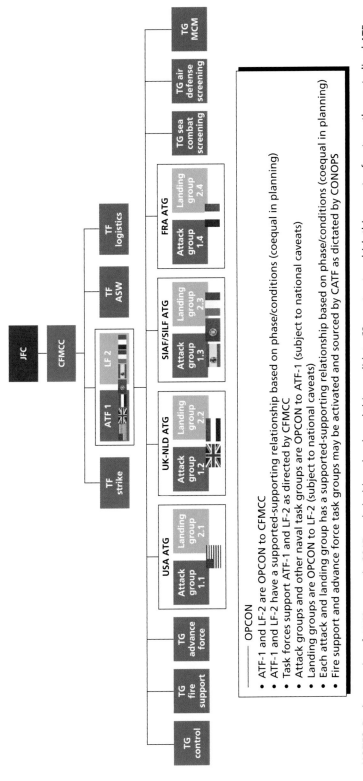

——— OPCON

- ATF-1 and LF-2 are OPCON to CFMCC
- ATF-1 and LF-2 have a supported-supporting relationship based on phase/conditions (coequal in planning)
- Task forces support ATF-1 and LF-2 as directed by CFMCC
- Attack groups and other naval task groups are OPCON to ATF-1 (subject to national caveats)
- Landing groups are OPCON to LF-2 (subject to national caveats)
- Each attack and landing group has a supported-supporting relationship based on phase/conditions (coequal in planning)
- Fire support and advance force task groups may be activated and sourced by CATF as dictated by CONOPS

NOTES: During a series of ALES events in 2017, stakeholders developed this candidate C2 construct—which this report refers to as the centralized ATF—for large-scale, multinational amphibious operations in an MJO+ conflict. The construct was further defined during an April 2018 ALES planner-level meeting in preparation for this wargame. Although a notional joint and maritime organization is depicted, the focus of the wargame was on the ATF and ATGs. LF = landing force; TF = task force; TG = task group.

2. Maritime support assets would be redirected from Andenes in the evening to Bodo and Tromsø in the morning hours as the operation proceeded.
3. The French ATG would be the reserve and needed to prepare to reinforce Bodo, attack Red's currently unlocated reserve force, or take Ostrov's distant island of Kamen (located 800 nautical miles northeast of Ostrov proper) if Red were to attack it.
4. There would be a single AOA; this would enable the CATF/CLF to quickly shift resources as dictated by the situation.
5. The CFMCC would be protecting the AOA from potential threats from the northwest and southwest.[2]

After the CATF and CLF established this CONOPS, a discussion of battlespace management ensued. However, aside from broad analogies to recent operational constructs and an intent to employ the CWC, specific authorities were not delineated for the more complex aspects of battlespace management, such as fires and airspace deconfliction.[3] The terms *sector*, *AO*, *AOA*, and *AOR* were used interchangeably and without what appeared to be a mutual understanding across the CATF/CLF staffs—this would, as the game went on, also increasingly complicate ATG efforts to develop supporting CONOPS.

To enable the CONOPS, the CATF proceeded to request additional assets and clarification regarding surface ship roles from the CFMCC, including

1. an appeal for additional antiair platforms to protect amphibious shipping and operational control (OPCON) over these assets
2. a request for OPCON of the *Queen Elizabeth* and *Cavour* carriers to conform with UKNLAF and SIAF/SILF's respective operating constructs in which amphibious ships would deploy in task groups with a carrier and operate with minimal restrictions on their employment
3. a confirmation that the *Queen Elizabeth* would be available in its amphibious configuration.

Additional CATF/CLF deliberations centered on the appropriate coordination mechanisms between the CATF/CLF and the CFMCC to include the delineation

---

[2] This assumption is critical to examine in future allied planning. The Naples Tabletop concluded that no naval organization is currently resourced to oversee a maritime campaign that includes large-scale amphibious operations and requires battlespace ownership of a significant portion of the theater.

[3] CWC allows assignment of commanders to specific warfare areas, such as air warfare or antisubmarine warfare, who in turn report to an overall commander who apportions assets and priorities among these commanders. Units may pass between different commanders as required by the tactical situation. CWC employs "command by negation," which means that subordinate commanders are empowered to carry out operations within mission orders up to the point that senior commander might negate them. CWC is intended to give maximum flexibility in assignment of assets and conduct of operations.

of boundaries and assurances that CATF/CLF efforts would receive adequate priority. There was particular concern that C2 seams might be present that would allow a theater-level threat to present a danger to the ATF. This concern was communicated to the CFMCC, but it was not completely clear that the coordination mechanisms were put in place to resolve CATF/CLF's concern.

The CATF/CLF also wanted to ensure that subordinate units were clear as to tasking and expressed concerns about the limited time provided to the ATGs between the issuing of the CATF/CLF statement of intent and the requirements for ATGs to generate a detailed plan. While some efficiencies were achieved by allowing a parallel planning path, the CATF and CLF were both concerned that there might not be time to develop the kinds of coordination mechanisms that might be required by NATO procedures.

Meanwhile, the ATGs began a parallel planning effort based on their respective assigned tasks. After submitting their initial CONOPS to the CATF/CLF and receiving additional planning guidance, each of the ATGs turned to a discussion of the advantages of maintaining a maximum level of control over integral forces; there was a strong attachment to national and habitual groupings and resistance to operating with a different mix of forces than what is currently routine. For example, the British and Dutch preferred to retain control over the air defense ships needed to protect their amphibious ships. Participants stressed that forces that train as a package during peacetime will perform best if they retain their constituent capabilities and operate under habitual authorities. There was, however, a reserved acceptance of the CATF and CLF needing flexibility to organize and reallocate assets based on the situation. The tension between the preference for routinized C2 structures and the need for adaptability would reappear once the landings commenced in Move 2.

At the CFMCC level, the discussion focused largely on which missions, functions, and resources should be delegated or assigned to the CATF/CLF and which should be retained by the CFMCC. There was immediate consensus on the principle that the CFMCC would delegate missions and functions that focused specifically on the ATF AOA to the CATF/CLF together with assets and resources that should be dedicated to those missions and functions. Missions and functions that focused beyond the ATF AOA would be retained by the CFMCC, along with the assets and resources needed for those missions and functions. CFMCC staff decided that they would retain and manage assets and resources for intelligence, surveillance, and reconnaissance; ASW; ballistic missile defense; Tomahawk land-attack cruise missiles; and air assets from the carriers.

In each case, the CFMCC would apportion some of these assets to the CATF/CLF as appropriate at a given time during the operation. Discussion focused on how air assets could be apportioned: Should the CFMCC apportion the carriers themselves (with their air wings) or only apportion the sorties generated by the carriers? Given the fluid nature of the operation, the additional mission to defend Kamen Island (retained by the CFMCC), and the on-call status to support operations on the European main-land, the CFMCC decided that retaining the carriers and allocating sorties provided a needed level of operational flexibility.

Move 2

To begin Move 2, RAND's control team provided a situational update. UKNLAF's landing at Andenes progressed quicker than anticipated as enemy forces were signifi-cantly lighter than estimated. The assault on Tromsø was disrupted by the downing of an MV-22 by man-portable air-defense systems, among other losses, and the U.S. landing force needed more time to accomplish its mission. Most concerning is that SIAF lost two Portuguese frigates to CDCM strikes while SILF encountered substan-tial resistance, remained blocked on the beach, and sought reinforcements to punch through Red's defenses.

Given the resistance in Bodo, the CATF/CLF directed SIAF/SILF to develop a revised scheme of maneuver and to plan on integrating the French reserve force to assist in breaking out from the landing. The ensuing discussion between SILF and the French landing force would present planning dilemmas that challenged the central-ized ATF concept, which is premised on the aggregation of integral—but not nec-essarily cross-compatible—elements into a sum greater than its parts. For example, while participants felt that placing French artillery and engineering elements under direct control of SILF (with NATO call-for-fire procedures and liaisons) would not present significant C2 issues, integrating maneuver forces that did not previously train together within the same geographic area would create a high degree of risk.[4] After a protracted discussion, participants proposed that French infantry would be staged in the northern half of Bodo, ready to move into the southern zone if SILF continued to experience heavy opposition. This course of action strongly suggests that the level of interoperability required for integrated infantry operations at the battalion and lower levels (i.e., modularity) does not exist within NATO's landing forces beyond the habit-ual relationships that are the basis for bilateral ATGs.[5]

Beyond Bodo, UKNLAF moved to the consolidation phase of its plan and began to consider handoff to security forces, reembarkation, and a range of possible taskings, such as sailing toward Kamen, which the French would no longer be able to cover with their focus on reinforcing Bodo. The U.S. ATG, while still heavily engaged in Tromsø, recognized and offered excess capacity to the CATF and CLF. The scenario's unfolding complexity reinforced the emerging consensus, first posited at Northwood, that an intermediate layer of command between the CFMCC and the ATGs would be essential to amphibious operations during an MJO+. This command, furthermore,

---

[4]   Beyond revealing a general concern for fratricide, the wargame (by design) did not include the tactical level of detail needed to identify exactly what could go wrong. A brief UKNLAF internal discussion about possibly retaking Kamen as part of a U.S.-UK task group surfaced similar concerns regarding landing force compatibility.

[5]   NATO's nonamphibious ground forces have similar characteristics and limitations.

requires multinational CATF and CLF staffs intimately familiar with the capabilities, equipment, tactics, and limitations of respective national capabilities. During this move, the CATF and CLF staffs assigned new missions to the subordinate ATGs, reallocated assets across the ATF, and coordinated complex defensive functions with the CFMCC—actions that would likely have been beyond the capacity of the CFMCC staff or independent ATGs to coordinate on their own without the intermediate level of command for amphibious operations.

CATF and CLF deliberations included renewed concern that open C2 seams had resulted in poor coordination mechanisms with the CFMCC. Neither the CFMCC nor the CATF/CLF apparently had sufficient forces available to counter the ASCM attack on the SIAF, which resulted in some damage but perhaps equally serious, a depletion of air defense ammunition. This shortfall resulted in replanning of overall task force maritime defense operations, recognizing that the CATF/CLF might thus be required to provide the bulk of task force defense. In addition, with the prospect of another amphibious operation on Kamen Island looming, the CATF/CLF noted that expanding operations outside the declared AOA would likely exceed their already stressed C2 capabilities. If a Kamen Island operation was required, they would recommend that the CFMCC create a separate amphibious task force or group for that mission and that the CFMCC exercise direct C2 over that element.

Given the unfolding battle in the AOA, the CFMCC discussion focused largely on four issues:

1. reconstituting and strengthening the air and missile defenses of the ATF
2. CFMCC's role in coordinating the salvage and repair of damaged ships
3. replenishment of ATF ships (especially with regard to ordnance)
4. who would lead a likely Kamen operation.

Regarding the first issue, the CFMCC adjusted the balance in apportioning air sorties for offensive counterair and defensive counterair operations and requested joint force air component commander offensive counterair support. Regarding the second and third issues, CFMCC decided that it was responsible for reloading, salvage, and repair and discussed the broad outlines of how CATF/CLF ships that were damaged or low on ordnance would be swapped with other CFMCC assets. CFMCC staff discussed the possibility of establishing rapid reload facilities in Norway and the United Kingdom, as well as options for ship repair in Norway, the United Kingdom, and France. On the last issue, there was a broad-ranging discussion of the advantages and disadvantages of tasking the Kamen operation to the CATF/CLF versus executing it at the CFMCC level, but no clear decision was reached.

Participant Observations and Recommendations

Caution must be shown in drawing conclusions from a single event. These insights are derived from execution of and deliberations during a tabletop-style wargame. In most cases, the recommendations require further investigation and additional testing through simulation or exercises before adoption. However, many of the participants coalesced around six observations and recommendations. Where divergent views were expressed, they are noted below.

Observation 1: *The centralized ATF facilitates appropriate division of labor and span of control between the CFMCC and amphibious forces.* The CFMCC in an MJO+ scenario would be challenged to oversee large-scale amphibious operations while simultaneously directing sea control operations in a large portion of the North Atlantic, defending the ATF from threats originating outside the AOA, coordinating joint functions with adjacent and higher commands, and facilitating maritime logistics.

- **Recommendation 1:** *Delineate CFMCC and ATF responsibilities.* Conduct allied staff planning to specify CFMCC-retained and ATF-delegated functions, responsibilities, and authorities during an MJO+ involving large-scale amphibious operations.

Observation 2: *The centralized ATF provides enhanced C2 capabilities and capacity that would likely be needed for large-scale multinational amphibious operations.* During the wargame, the ATF was effective in allocating resources and managing complex, cross domain operations across a large battlespace involving multiple ATGs and objectives. While not appropriate to every situation, this C2 construct would be a strong candidate for MJO+ scenarios involving large-scale amphibious operations.

- **Recommendation 2a:** *Test the centralized ATF in future exercises.* Although the wargame demonstrated the centralized ATF's potential, and previous ALES events discussed its viability compared with competing constructs, exercises are required to validate and evolve the construct. SACEUR's Annual Guidance on Education, Training, Exercises, and Evaluation presents an appropriate mechanism to register this need.
- **Recommendation 2b:** *Develop a NATO concept paper for amphibious operations at the MJO+ level.* The concept paper could be used as the basis for experimentation, exercise development, and eventual revisions to doctrine.

Observation 3: *The centralized ATF construct requires commanders and staff officers with experience in multinational operations and expertise in amphibious warfare.* As envisioned by ALES participants in this wargame and prior events, coordination with the CFMCC and effective oversight of multiple ATGs requires that the CATF and CLF have the requisite rank and expertise and are provided multinational staffs of the size and composition beyond recent precedent. Defining manning for the centralized ATF could inform deliberations about whether it should be standing, rotational, or sourced at time of incident.

- **Recommendation 3:** *Specify requirements for sourcing the CATF, the CLF, and their multinational staffs.* Conduct allied staff planning to determine the number and qualifications of personnel needed for the centralized ATF.

Observation 4: *National and bilateral ATGs are designed and generally employed as integral elements, but larger-scale operations would benefit from additional interoperability and flexibility.* ATG staffs were uncomfortable with aspects of operating under a construct that differed from their previous experiences. Concerns included ceding authorities typically vested in a national or bilateral CATF and CLF to the ATF level, detaching ships to operate as part of a pooled ATF or CFMCC asset group, and operating within the same ground battlespace as another ATG. Participants at this and previous ALES events felt that national and bilateral ATGs are most effective when employed as integral units and expressed reservations about the political implications and tactical risk inherent in mixing and matching forces (particularly for ground maneuver). Despite these reservations, many participants recognized that an ATF could encounter situations where it would need to shift resources rapidly; this includes the possibility of assigning an ATG, or a subset of an ATG, in support or under the temporary control of an adjacent force it may not have trained with. They acknowledged that the issue is underexplored, given that MJO+ is only a recent focus area for NATO planning.

- **Recommendation 4:** *Identify and pursue tactical interoperability needs beyond current habitual relationships.* However, any new commitments to pursue interoperability and exercise ATG-to-ATG support relationships should be undertaken with a realization that the proficiency of national and bilateral ATGs as integral force elements will remain critical and requires continual attention.

Observation 5: *There is an uneven understanding of maritime and amphibious doctrine regarding C2-related issues, particularly battlespace management.* At several points during the wargame, participants came away from synchronization meetings with differing views of how the ATF planned to divide and control the battlespace. Misperceptions were especially prominent between the CATF/CLF and leadership of the various ATGs. Two factors may explain this observation. First, some participants acknowledged that they defaulted to using concepts and terminology based on their recent exercise and operational experiences rather than established NATO doctrine. Second, given that ALES arrived at the centralized ATF concept only recently, existing doctrine may not be sufficiently clear regarding the more complex aspects of C2 such as airspace control and fires integration. The wargame's compressed nature prevented fully considering the misunderstandings or demonstrating potential consequences, but participants acknowledged the need to study doctrinal C2 constructs and their application in high-end amphibious operations.

- **Recommendation 5:** *Enhance understanding and employment of amphibious battlespace management doctrine across NATO's amphibious forces.* Conduct workshops and other planning events to train staff officers in allied maritime and amphibious doctrine, including airspace control and fires integration.

Observation 6: *Ongoing initiatives such as NATO Command Structure Adaptation, the Allied Maritime Posture initiative, and the NATO Readiness Initiative provide an opportunity to operationalize the centralized ATF concept.*[6] During the final plenary, participants expressed that the increased profile of maritime forces in the NATO political-strategic landscape and the potential value of the centralized ATF capability could enable and energize detailed planning for further evolution of amphibious forces by national, multinational, and NATO organizations.

- **Recommendation 6:** *Develop a NATO road map for generating and employing the centralized ATF.* Key work strands could include planning and sequencing exercises, assessing CIS interoperability, specifying the ATF's staff structure, refreshing NATO maritime and amphibious doctrine, and enhancing the role of amphibious forces in current and emerging allied operational plans.

Chapter Summary

This wargame surfaced the potential benefits of the centralized ATF while also pointing to actions needed for planning, testing, and implementing this C2 construct. Participants—many of today's maritime and amphibious leaders within the Alliance—came away with a clearer understanding of the challenges they face in preparing their forces to integrate inside what is an evolving NATO command structure. More broadly, there is a recognition that while NATO has recently taken steps to enhance its preparedness for an MJO+, additional work remains in the amphibious realm to harness significant

---

[6]   NATO Command Structure Adaptation is an initiative that includes the establishment of JFC Norfolk to oversee potential operations in the North Atlantic. The Allied Maritime Posture initiative is intended to inform allied investment and capability development for maritime forces, including amphibious forces. The NATO Readiness Initiative aims to enhance the readiness of existing national forces and their ability to move within Europe and across the Atlantic; allies have committed, by 2020, to having 30 battalions, 30 air squadrons, and 30 naval combat vessels ready to use within 30 days.

national and bilateral capacity into a coherent, integrated multinational capability in the form of a centralized ATF.

A final caveat is in order. While the wargame illustrated that the centralized ATF is an attractive option for organizing and employing NATO's amphibious forces, **no C2 construct can be successful without sufficiently ready and capable forces**. In this regard, properly resourcing amphibious forces and tailoring and sequencing a collective and progressive exercise schedule—to include national and bilateral events—are perhaps the most important considerations for NATO in building a potent and flexible amphibious capability for the Alliance.

CHAPTER SIX

Conclusion

Over the past two years, U.S. and European amphibious leaders formed ALES, identified amphibious C2 for MJO+ as a priority area, and arrived at a tentative solution for C2 in the form of the centralized ATF, albeit with the caveats that NATO's broader maritime structure remains a work in progress and the broadly agreed-upon amphibious C2 construct is a baseline that will need further refinement and testing. Recent NAC documents call for a renewed focus on high-end maritime capabilities and have stimulated review and transformation of NATO's maritime C2 for large-scale operations.

Having examined the progression of ALES through the outcomes of each individual event, this chapter presents holistic findings concerning the evolution of maritime and amphibious C2; these findings are derived from the totality of the three events as well as additional interviews with concept developers and practitioners. The report concludes by summarizing ALES's progress and offering potential next steps for NATO.

Findings from ALES Research

Finding 1: *ALES nations have considerable amphibious capacity, but these forces have been an underrecognized asset in NATO.* Individual national and bilateral ATGs currently form the core of allied amphibious forces. In the case of bilateral entities such as the UKNLAF and SIAF/SILF, habitual training and deployments over the course of decades have resulted in integral force packages with a level of uniformity approaching that of purely national units such as the French and U.S. task groups. These forces, while capable of conducting battalion or brigade-size operations, most frequently train and are employed below the battalion level. Further, current NATO force planning and crisis response structures may not take advantage of the full potential of available amphibious capabilities.

As NATO responds to a changing security environment and moves to improve its maritime capabilities, amphibious forces offer a flexible and potent instrument that can enhance deterrence through early deployment and project credible combat power from the maritime domain. Allied ATGs maintain the requisite shipping, connectors, and landing forces to conduct multibrigade operations. U.S. forces, in the form of an ESG and MEB, are the most capable in penetrating A2/AD environments, but European amphibious formations, when enabled by joint forces, could be employed in a range of scenarios including amphibious demonstrations, raids, subsidiary landings, and assaults against properly prepared objective areas. Energized by ALES, NATO is now considering how to transform the sum of these amphibious elements into something greater than its parts.

Finding 2: *The centralized ATF construct developed by ALES stakeholders offers a mechanism to leverage NATO's amphibious capacity by aggregating national and bilateral capabilities into a coherent C2 structure.* There is an emerging consensus around a baseline C2 structure—the centralized ATF—for amphibious operations in an MJO+ scenario. Maritime and amphibious leaders recognize that this construct requires testing and validation in exercises and is not appropriate for every situation.

Wargaming enabled discovery of three key challenges to the centralized ATF. First, there is a need to delineate roles and responsibilities for a CFMCC and centralized ATF. The centralized ATF concept could enable a CFMCC to concentrate on the broader maritime campaign, but there are outstanding questions regarding prioritization, allocation, management, sequencing, and control of assets as amphibious operations unfold: What are the arrangements for protecting the AOA from external threats? Who leads information operations? Answering these types of questions (and others documented in Appendix C) and developing options requires planning and continuous refinement as NATO's maritime C2 structure evolves.

Second, even with CFMCC-ATF roles defined, employing a multinational ATF at echelons above a single brigade or BLT poses new challenges for U.S. and allied amphibious forces. Complex aspects of battlespace management such as integrated air and missile defense or cross-boundary fires integration will require selection and refinement of C2 procedures that provide an appropriate blend of centralization and delegation of authorities while retaining operational flexibility in a potentially degraded communications environment.

Finally, in task organizing, commanders will need to balance the flexibility to tailor force groupings afforded under a centralized ATF against the stability and continuity provided by employing national and bilateral forces as integral elements. Participants at ALES events emphasized that national and bilateral ATGs are most effective when employed as integral units and expressed reservations about the political and tactical risk inherent in mixing and matching forces, particularly ground maneuver elements. Many amphibious leaders nevertheless recognize that a multinational ATF could encounter situations where it would need to shift resources rapidly; this includes the possibility of assigning an ATG, or a subset of an ATG, in support or under the temporary control of an adjacent force it may not have trained with. Such relationships could be exceptionally difficult to arrange during a contingency, making rehearsals in peacetime essential.

Additionally, tactical wargames (potentially with simulations) and exercises offer appropriate venues to further test the centralized ATF construct; robust flag and general officer participation will be important in developing responses to these complex and potentially politically sensitive issues.

Finding 3: *The centralized ATF construct requires commanders and staff with experience in multinational operations and expertise in amphibious warfare.* ALES events demonstrated that the commanders of the ATF will be required to exercise broad, complex C2 functions that include coordination with the CFMCC and effective oversight of multiple ATGs. Individuals filling the roles of CATF and CLF must have the requisite rank and experience to perform these functions. In addition, they should have multinational staffs of sufficient size and expertise demanded by the scope and scale of the amphibious force and its operations.

- For the CATF, there is no clear source of a corresponding navy officer and core staff. The U.S. ESG is a potential source for this capability but would require augmentation and training. The recent reconstitution of the U.S. 2nd Fleet may offer additional options.
- For the CLF role, the United States is best positioned to provide a two- or three-star officer and core staff from within II Marine Expeditionary Force.
- The size of a multinational staff could be as high as 500 personnel, depending on composition of the ATF and division of functions between the ATF and the CFMCC.
- Given the C2 and CIS functionality required and the potential size of the ATF staff, the choice of an ATF flagship is an important planning factor. While U.S. platforms have frequently served as flagships for larger NATO maritime elements, several European ships appear to have potential to serve in this capacity.

Allied staff planners could define manning requirements for the centralized ATF to inform deliberations about CLF and CATF selection, staff composition, flagship selection, and whether the organization should be standing, rotational, or sourced at time of incident.

Finding 4: *Knowledge and experience in large-scale amphibious operations has atrophied across allied naval and landing force practitioners.* Commanders and staffs should reexamine and reinforce maritime and amphibious doctrine for large-scale operations. Throughout ALES discussions, participants often had diverging perspectives on terminology and concepts. This may be a result of participants defaulting to concepts and terminology based on recent exercise and operational experiences rather than established NATO doctrine. Another possible explanation is that existing doctrine may not be sufficiently clear regarding its application to operations of a multinational force above the brigade level. Regardless, allies should stress the development of broader amphibious expertise in training and exercises as well as invest efforts in evolving doctrinal and operational concepts for the conduct of amphibious operations against new and emerging threats.

Potential Next Steps for NATO

ALES's focus on maritime and amphibious C2 has identified key issues, surfaced potential solutions, and stimulated action within national, bilateral, multinational, and Alliance organizations. Seminars and wargames provided allied military leaders and their staffs with a structured process through which to design and explore alternative C2 constructs for amphibious operations. Figure 6.1 depicts the progression of design, wargaming, and exercising needed to arrive at a construct optimized for large-scale NATO operations.

With the initial rounds of C2-centric wargaming at the operational level completed, a next step is command post exercises (CPXs) and live exercises that include a multinational ATF, to include its command ship.[1] Events such as CPXs enable the creation or refinement of templates, orders, and standard operating procedures that amphibious leaders acknowledge need attention for multi-ATG operations under a centralized ATF. Exercises could also permit testing issues that are difficult to wargame but directly affect C2, such as CIS architecture and performance.

Figure 6.2 summarizes progress since 2016 and offers potential next steps for NATO. Although exercises present the most immediate mechanism to improve operational proficiency, generating a full allied ATF capability requires a broader set of actions. Realizing a NATO ATF will require that the Alliance's political commit-

Figure 6.1
Steps for Exploring Command and Control as of Mid-2018

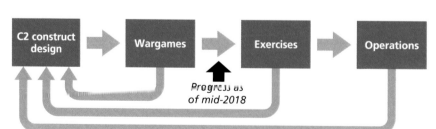

---

[1] Exercises could be supplemented with tactical-level wargaming and modeling and simulation, which would be needed to explore the relative strengths and weaknesses of each ATG and inform expectations for most likely support relationships and reinforcement plans.

Figure 6.2
ALES Progress and Potential Next Steps for NATO

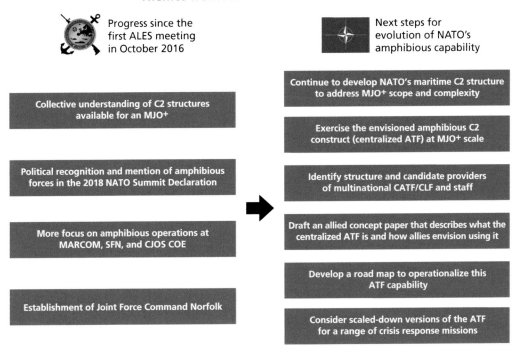

ment to amphibious operations be reinforced with allied planning, national resourcing, and military implementation. The following six steps could form the core of a NATO strategy to evolve its amphibious capability:

1. Continue to develop NATO's maritime C2 structure to address MJO+ scope and complexity.
2. Exercise the envisioned amphibious C2 construct (centralized ATF) at MJO+ scale.
3. Identify structure and candidate providers of multinational CATF/CLF and staff.
4. Draft an allied concept paper that describes what the centralized ATF is and how allies envision using it.
5. Develop a road map to operationalize the ATF capability.
6. Consider scaled-down versions of the ATF for a range of crisis response scenarios.[2]

---

[2] This could include examining the role of amphibious forces in the NATO Response Force and Standing NATO Maritime Groups.

Other issues and questions across the DOTMLPF-I spectrum that merit further analysis, additional consideration in planning, or testing in exercises are identified in Appendix C.

Throughout wargames and seminars, participants noted that MARCOM, as SACEUR's maritime adviser, is uniquely positioned to work with ALES stakeholders (and other NATO entities) to pursue this or a similar agenda. For example, MARCOM could advise SACEUR on how amphibious forces could be integrated into NATO's Readiness Initiative or inform the revision of graduated response plans in close consultation with allies that possess amphibious capabilities.[3]

---

[3]   Allies have committed, by 2020, to having 30 battalions, 30 air squadrons, and 30 naval combat vessels ready to use within 30 days.

Command and Control Constructs Explored

Maritime Command and Control

Figure A.1 shows the "as is" C2 construct employed on the first day of the Naples Tabletop. It represents what ALES stakeholders believed at that time (spring–summer 2017) to be a possible arrangement for high-end amphibious operations. The alternative construct shown in Figure A.2 was designed as a potential starting point for revision of the "as is" construct. The intent was to explore a divided JFC-SFN construct on Day 1, with participants incorporating supported-supporting relationships into their CONOPS. The second day would test a unified construct under a single CFMCC.

Figure A.1
Baseline Naples Tabletop Exercise Command and Control Construct

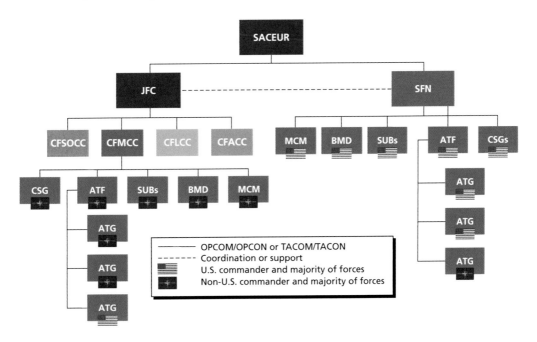

Figure A.2
Alternative Naples Tabletop Exercise Command and Control Construct

```
                        ┌─────────┐
                        │ SACEUR  │
                        └────┬────┘
                        ┌────┴────┐
                        │   JFC   │
                        └────┬────┘
        ┌──────────┬─────────┼──────────┬──────────┐
   ┌────┴────┐ ┌───┴───┐ ┌───┴───┐  ┌───┴───┐
   │ CFSOCC  │ │ CFMCC │ │ CFLCC │  │ CFACC │
   └─────────┘ └───┬───┘ └───────┘  └───────┘
          ┌────┬───┼─────┬───────┬────────┐
      ┌───┴─┐┌─┴──┐  ┌───┴──┐ ┌──┴──┐ ┌──┴──┐
      │CSGs ││ATF │  │ SUBs │ │ BMD │ │ MCM │
      └─────┘└────┘  └──────┘ └─────┘ └─────┘
          ┌──┴──┐
          │ ATG │
          └─────┘
          ┌─────┐
          │ ATG │      ──── OPCOM/OPCON or
          └─────┘           TACOM/TACON
          ┌─────┐
          │ ATG │
          └─────┘
```

Amphibious Command and Control

Figure A.3 shows the centralized ATF construct as envisioned in Northwood and later refined in the lead-up to the Stavanger Wargame. Figure A.4 shows a more detailed version of the same construct. The centralized ATF is designed to "plug into" a maritime C2 construct of some form.

Figure A.3
Simplified Centralized Amphibious Task Force Construct

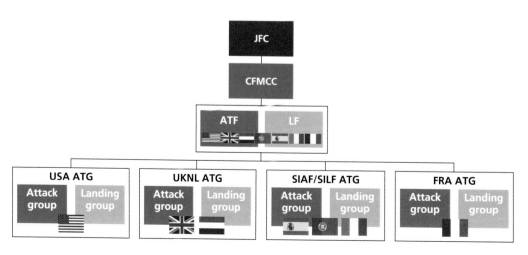

Figure A.4
Stavanger Wargame Blue Command and Control Construct

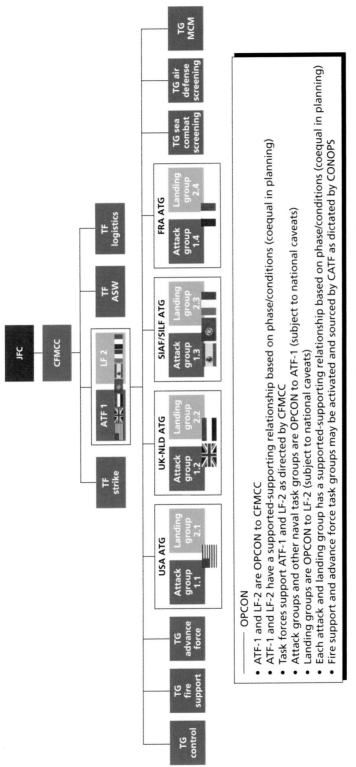

NOTE: During a series of ALES events in 2017, stakeholders developed this candidate C2 construct—which this report refers to as the centralized ATF—for large-scale, multinational amphibious operations in an MJO+ conflict. The construct was further defined during an April 2018 ALES planner-level meeting in preparation for this wargame. Although a notional joint and maritime organization is depicted, the focus of the wargame was on the ATF and ATGs. LF = landing force; TF = task force; TG = task group.

Scenarios and Forces

The Naples Tabletop

Scenario and Red Forces

This TTX is set in late summer 2018. A period of escalating tensions led to a Red attack against the Baltic states. Red struck in June, surprising the Alliance by launching major invasions against Estonia and Latvia. NATO declared Article 5 of the NATO Treaty, with the NAC issuing an execution directive for the appropriate contingency plans. The deployment of the Alliance's rapid response forces to Estonia, Latvia, Lithuania, and Poland began a few days prior to hostilities.

Red anticipated a quick and decisive victory against weak NATO opposition. However, national defense forces, supported by in-place Enhanced Forward Presence battle groups, early arriving rapid response force units, and other units deployed by the United States and other NATO members resisted the attack. NATO's air forces launched waves of attacks into Red's sophisticated air defenses, suffering losses but inflicting heavy attrition on the invading units and their supporting elements. At the end of June, NATO forces under the command of JFC Brunssum tenuously held the line against continuing Red pressure.

Both sides then focused on reinforcing their respective positions, marking a potential battle for the Atlantic as the next critical phase of the conflict. To enhance interdiction of NATO's vital transatlantic SLOC, Red seized the lightly defended Norwegian airbases at Andenes, Bodø, and Tromsø by paratroop drops. Red then deployed coastal defense cruise missiles, long-range surface-to-air missiles, and other assets to extend its A2/AD envelope deeper into the Norwegian Sea. In doing this, Red intended to support the movement of both surface and subsurface assets from its Northern Fleet against NATO's Atlantic SLOC and enhance protection of its strategic submarines in their Arctic Sea bastions.

The assessed Red order of battle in Norway is depicted in Figure B.1.

Both sides sought to limit the war to the conventional level, and also to prevent its geographic expansion beyond northern and western Europe. NATO conducted only limited tactical strikes into Red territory, primarily against air defense targets. Red

Figure B.1
Naples Tabletop Red Force List and Objectives

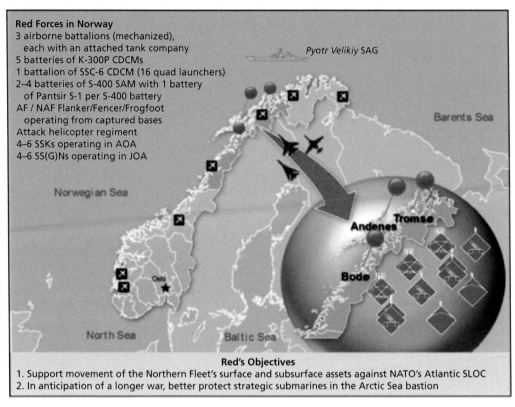

Red Forces in Norway
3 airborne battalions (mechanized),
  each with an attached tank company
5 batteries of K-300P CDCMs
1 battalion of SSC-6 CDCM (16 quad launchers)
2–4 batteries of S-400 SAM with 1 battery
  of Pantsir S-1 per S-400 battery
AF / NAF Flanker/Fencer/Frogfoot
  operating from captured bases
Attack helicopter regiment
4–6 SSKs operating in AOA
4–6 SS(G)Ns operating in JOA

Pyotr Velikiy SAG

Barents Sea

Norwegian Sea

Tromsø
Andenes
Bodø

North Sea

Baltic Sea

Oslo

Red's Objectives
1. Support movement of the Northern Fleet's surface and subsurface assets against NATO's Atlantic SLOC
2. In anticipation of a longer war, better protect strategic submarines in the Arctic Sea bastion

NOTE: CDCMs = coastal defense cruise missiles; SSKs = attack submarine; SSGNs = guided-missile submarines.

(with the exception of the Norway operation) limited its attacks to the territories of the Baltic States and Poland. In the southern region, Red sent submarines and surface vessels into the Mediterranean Sea before the onset of hostilities, using bases in Syria and the Black Sea for support. Red aviation conducted reconnaissance against Bulgaria, Romania, and Turkey, and Long-Range Aviation flew missions over the Black Sea region and the Mediterranean, but there were no attacks against NATO forces in the southern region.

To counter Red's move, SACEUR directed SFN to assume responsibility for securing the SLOC east of 45° west longitude while ordering MARCOM to establish a CFMCC with the mission of degrading Red A2/AD capabilities within the JOA and conducting an amphibious landing to retake the three captured airfields to remove the Red A2/AD threat from Norwegian soil. SFN is tasked with providing support to the JFC for this operation. JFC and SFN are ordered to develop a CONOPS and submit a coordinated C2 construct and measures to SACEUR with recommendations for

- support and C2 relationships between JFC and SFN
- assignment of tasks and allocation of forces
- designation of a CATF/CLF
- adjustments to battlespace geometry.

All NATO members agreed to provide the basing, overflight, and transit permissions required by SACEUR. No non-NATO nations are participating in the naval campaign.

Blue Forces

Table B.1 lists the ships, submarines, maritime patrol aircraft, and landing forces designated by game controllers as operationally available. Three assumptions dictated what assets would be operationally available:

1. The United States would provide three CSGs, shipping for a MEB afloat, and an additional MEU or independent amphibious element.
2. Remaining TTX nations would provide between one-half and three-fourths of their navies for force generation and employment.[1]

Table B.1
Naples Tabletop Summary of Available Assets

Asset	Total Operationally Available for TTX	ESP	FRA	GBR	ITA	NLD	NOR	PRT	USA	Other NATO
Aircraft carriers	5	—	1	—	1	—	—	—	3	—
Amphibious ships	27	2	2	4	2	2	—	—	15	—
Major surface combatants	68	6	10	10	9	3	3	3	17	7
Mine warfare ships	40	3	6	8	5	3	4	—	2	9
Attack submarines	29	2	3	4	4	2	3	1	8	2
Maritime patrol aircraft	52	—	3	—	—	13	3	—	24	9
Battalion landing teams	9.5	1	1	2	1	1	—	0.5	3	—

---

[1] This aggressive assumption made a large force available for the examination of C2 in a large-scale operation.

3. In addition to the TTX participants, Belgium, Canada, and Spain would be able to participate in this operation based on TTX controller-provided assumptions. Other NATO navies were committed elsewhere in the theater.

A draft version of the table in B.1 was provided to participants as a starting point for sourcing. Several national contingents identified assets in addition to the ones listed in the table that could be made available for such a scenario:

- The United Kingdom modified the number of BLTs available from one to two.
- Portugal offered that it has one-half of a BLT equivalent to contribute as a standalone or integrated element into a multinational ATG.
- France modified the number of operationally available maritime patrol aircraft to three.
- The United States clarified to other participants that the MEB includes a Marine aircraft group.
- Nations made minor adjustments to major surface combatants based on recent or expected operations, maintenance cycles, and procurements.

The Stavanger Wargame

Scenario and Red Forces

This wargame is set in May–June 2019. A period of escalating tensions led to a Red invasion of the Baltic States in early May. NATO declared Article 5 of the NATO Treaty and activated contingency plans that began the deployment of the Alliance's rapid response forces to Estonia, Latvia, Lithuania, and Poland. While both sides initially attempted to confine hostilities to the immediate environs of the Baltic region, the conflict escalated to include limited operations across Europe.

In the southern region, Red sortied submarines and surface vessels into the Mediterranean Sea prior to the onset of hostilities, using bases in Syria for support. Red aviation conducted reconnaissance against Bulgaria, Romania, and Turkey. Long-Range Aviation flew missions over the Black Sea region and the Mediterranean, but there were no attacks against NATO forces in the southern region.

Red deployed major elements of its surface, subsurface, and naval aviation forces to protect the maritime approaches to its homeland. In the north, Red dispersed surface components of its Baltic Sea Fleet, including its amphibious forces, to reinforce its Northern Fleet. These steps were intended to interdict NATO air and naval attacks on the homeland, protect Red strategic submarines in their bastions, and interfere with NATO efforts to deploy forces into theater.

In support of these operations, in mid-June, Red seized three air bases on the NATO member island state of Ostrov and began deploying long-range air defense and coastal defense cruise missiles into the lodgments surrounding them.

Red forces in Ostrov consist of naval infantry battalion tactical groups, long-range air defense and coastal defense cruise missile units, and rotating fighter/attack aircraft detachments. These formations are drawing on Ostrov's resources to supplement organic logistics and sustain themselves.

The Combined Force Maritime Component Command–Northeast (CFMCC-NE) has responsibility for air and naval defense of NATO's northern flank and prosecuting maritime operations against Red forces in its area of operations. In response to the attack on Ostrov, the CFMCC-NE and its subordinate Amphibious Force–Northeast are tasked to defeat Red forces on the island, retake the seized air fields, and hand off control to a follow-on NATO security force. Table B.2 shows the wargame's geography while Figure B.3 provides an overview of available forces for the operation.

Figure B.2
Stavanger Wargame Geography

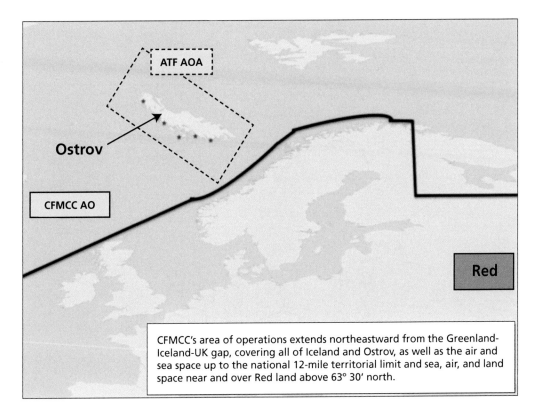

Blue Forces

Table B.2
Stavanger Wargame Summary of Available Assets

Asset	Total Available	ESP	FRA	GBR	ITA	NLD	NOR	PRT	USA	DNK
Aircraft carriers	5	—	1	1	1	—	—	—	2	—
Amphibious ships	24	2	2	3	2	2	—	—	13	—
Cruisers and destroyers	26	—	4	4	3	—	—	—	15	—
Frigates	24	5	4	4	4	5	—	2	—	—
Mine counter-measures ships	14	—	—	4	—	2	3	—	3	2
Battalion landing teams or equivalents	10	1	2	2	1	1	—	—	3	—

NOTE: Does not include Portuguese contribution to the Spanish-Italian ATG.

Figure B.3
Stavanger Wargame Blue Task Organization

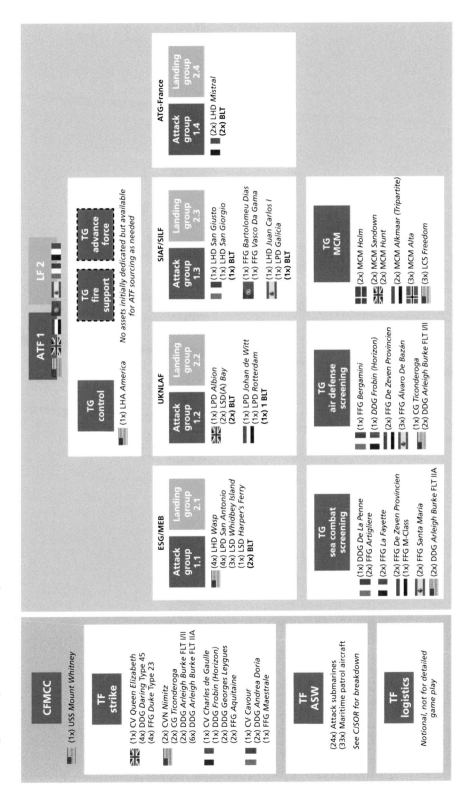

APPENDIX C

DOTMLPF-I Summary

This appendix documents issues and questions across the DOTMLPF-I spectrum that event participants identified as meriting further analysis, additional consideration in planning, or testing in exercises. Some of the topics listed may have been resolved by individual services, nations, or allied working groups following the Stavanger Wargame, while others require longer-term planning and further dialogue among NATO's amphibious leaders and staffs. For issues where participants expressed diverging viewpoints, we note the contending perspectives.

Doctrine

Mutual Understanding of C2 Principles and Definitions

Discussions during ALES events revealed that participants at times held varied and incompatible perceptions of doctrinal principles and terms related to command relationships and battlespace management. NATO and national doctrines do address fundamental concepts. However, participants consistently sought clarification on what they believed to be the implications of those concepts, such as the distinction between an ATG being assigned an AOA designation versus an AO (area of operations) designation.

- How can doctrinal fundamentals be better communicated to promote a mutual understanding across allied amphibious leaders and staffs?
- Is there agreement between national and NATO doctrines on fundamental concepts and terminology?
- Does NATO doctrine need updating to establish a more universal understanding of doctrinal principles?
- Why are differences in understanding and expectations arising?
- Which entities are responsible for establishing and educating amphibious leaders and staffs on the more complex aspects of doctrine?

The Evolving Threat Environment

Enemy A2/AD capabilities envisioned in MJO+ operations introduce novel and sophisticated threats to NATO's amphibious forces. Allied amphibious doctrine provides a common philosophy and approach for planning and operations, but wargame and seminar participants had an uneven understanding of how current doctrine could be applied to the operating environment of an MJO+.

- Do allied maritime concepts, doctrine, and operational procedures need updates to account for more capable potential adversaries and the challenges of an A2/AD environment?
- What options exist for conducting C2 of the ASW battle as a coordinated effort among the CFMCC, MARCOM, SFN, and the theater ASW commander?
- How should amphibious forces integrate with existing and future headquarters organizations on land?

Organization

Disaggregation of National or Bilateral Habitual Formations

Several tactical scenarios led participants to disaggregate and reorganize what are typically integral force packages. Some participants raised concerns about the risks associated with such mixing and matching of forces. The ability to reorganize at sea—let alone when in contact—and reassign individual ships or landing force elements from one ATG to another was questioned, with some players inquiring whether the CATF and CLF expected an unrealistic degree of scalability and flexibility.

For example, some ships and units are designed as multipurpose platforms; removing them from a battle group to selectively use one of their capabilities, such as air defense, could degrade their other functions, such as the C2 of their original formation. Disaggregating battle group formations can also strain sustainment by separating assets and units from their previously organic logistics. Other participants were apprehensive about conducting operations after portions of their traditionally organic ATG, such as submarines or surface combatants, had been removed from their standing national force package.

- How should NATO employ ATGs to maximize flexibility without degrading purpose-developed capability mission packages?
- What are the realistic limitations to scalability and flexibility for dynamically organizing multinational forces?
- Does NATO's amphibious doctrine, training, and education need to be expanded to incorporate guidelines for compositing or disaggregating allied ATGs?

- At what level should ATGs train to be disaggregated from their planned original composition?

For additional discussion regarding habitual relationships, see the "Interoperability" section below.

Organizational Impacts on Battlespace Management

As discussed in the "Doctrine" section above, a debate developed during the wargame over terminology referencing battlespace responsibility. Some participants assumed that ATGs were being assigned an AOA, with all ATF elements placed under TACON of the ATG, while others understood that the ATGs were assigned an AO, in which some task force elements remained under the full control of the ATF and/or the CFMCC. During discussion, participants offered that assignment of individual AOAs to each ATG could better satisfy principles of unity of command while reducing complexity and friction. Others felt that the individual ATGs may not have the expertise and capacity to exercise battlespace management of all the functions required by an AOA.

- What are the criteria for determining the appropriate battlespace geometry (AO versus AOA) for an ATG?
- What are the implications for targeting of different battlespace management and organizational approaches?
- What kind of C2 relationships are appropriate for managing capabilities that operate outside of an ATG's AO but have effects within it, such as air and missile defense or long-range strike platforms?

CATF and CLF Staffs

Envisioned scenarios could require a CATF and CLF staff capable of complex battlespace management and coordination of coalition forces across multiple domains and functions, including cyberspace and other information realms. Some participants noted that many existing staff structures (e.g., the U.S. ESG-2 command element or even larger U.S. staffs) lack the capacity, education, and training that may be required to exercise the full spectrum of C2 functions for the centralized ATF.

- What are the manning requirements for a multinational CATF/CLF headquarters organization?
- What changes would need to be made to the ESG or a NATO HRF(M)'s headquarters organizational structure, capacity, and expertise if it were to be designated as the CATF for an MJO+ amphibious operation?
- Should a standing, multinational CATF/CLF headquarters be established, or can existing structures be aggregated when needed to satisfy staff requirements?

NATO Command Structure Adaptation

NATO recently approved the creation of JFC Norfolk to manage operations in the North Atlantic region, which will operate alongside a reconstituted U.S. 2nd Fleet. JFC, MARCOM, and SFN roles within the NATO maritime command structure also continue to evolve.

- How does NATO Command Structure Adaption affect plans for amphibious C2?
- How could the centralized ATF "plug into" the emerging structure?
- Should a new or existing maritime command (JFC Norfolk, MARCOM, SFN, the U.S. 2nd or 6th Fleets, or other HRF[M]) be designated as a CFMCC or ATF/CATF in NATO plans?

Training

Realistic Exercises to Confirm Assumptions and Refine Concepts

Enemy capabilities have increased in range, lethality, and sophistication. NATO capabilities have similarly improved but involve technically challenging C2 arrangements not yet explored in ALES events. Fully appreciating the level of complexity represented by these changes will require large-scale realistic training exercises.

- How can exercises be redesigned and expanded to test C2 constructs against a high-end threat?
- How should exercises be scoped to focus on MJO+ scenarios?
- How can exercises better focus on coordination across ATGs?

Allied Experience with Amphibious Operations

ALES exercises highlighted the operational necessity of scalable interoperability among allied amphibious capabilities. However, participants noted that many of their forces lacked recent exercise or operational experience demonstrating the anticipated degree of integration, with some exceptions for existing habitual bilateral relationships. A common theme was the desire to improve this capability without detriment to established habituality.

- How can future exercises be tailored to better promote and test multinational integration during amphibious operations?
- How can training maximize the familiarization of allied nations' capabilities and limitations among NATO officials?
- Can training evaluation criteria be incorporated to assess nations' ability to share objective areas, deconflict fires, provide effects in support of each other, and the like?

Materiel

Communications and Information Systems

ALES exercises did not explore system-level interoperability capabilities and limitations of current communications architectures. The ability to share and disseminate classified information was also not explored in detail. Combining multinational forces under various C2 structures requires the ability to agilely employ multiple, redundant forms of communication. Further, the future threat environment will require strict adherence to various levels of emissions control.

- What are the minimum CIS needs for components of a NATO ATF, and what are the current shortfalls in CIS equipment used by NATO's amphibious forces?
- Do national or allied policies (e.g., restrictions on installation of U.S. classified systems) inhibit exercise of CIS functionality in envisioned scenarios and plans?
- Do allied amphibious forces have a common set of signature management and emissions control practices?
- Do NATO plans fully consider use of European platforms, some of which have made significant upgrades to their communication suites?

Sustainment

ALES events focused on operational-level concepts of employment, force allocation, and C2 constructs, and the flexibility and scalability of allied nations' capabilities and units was emphasized to address operational and C2 considerations. From a logistics perspective, participants noted that the ability to sustain disaggregated national force packages, as well as, potentially, elements of other allied nations must be explored further because logistics represents an area that does not scale easily.[1] This issue may be particularly acute for an ATF's aviation elements.

- What sustainment factors will be most stressed under a centralized ATF construct?
- What adjustments to current and future sustainment practices can be made to maximize the ability for various parts of the ATF to support one another?
- How can ATF task organization and amphibious ship embarkation plans maximize flexibility without sacrificing sustainment capabilities?
- How can multinational logistics capabilities be tested during future exercises?

For a discussion of classes of supply, see the "Interoperability" section below.

---

[1] For example, while a MEB has a notional ability to sustain its operations for 30 days, once a BLT is placed ashore and the MEB's remaining forces are directed to support another objective, the BLT may be incapable of sustainment beyond several days.

Leadership and Education

Knowledge of Multinational Capabilities

Discussions among ALES exercise participants often included ad hoc lessons on national and service capabilities possessed by respective allies' amphibious and maritime forces. An accurate understanding of allies' national force capabilities and limitations would benefit allied planning, as well as dynamic force employment at time of crisis.

- Do NATO military officials have an adequate understanding of allied nations' amphibious capabilities and limitations to conduct proper planning?
- Do allies have a mutual understanding of the HRF(M) headquarters construct, its capabilities, and its limitations?
- How could the understanding of NATO allies' amphibious capabilities and limitations be improved?
- Does NATO have an amphibious planners' guide or similar document?

Institutional Knowledge Among Senior Leaders

Senior leaders with assignments to NATO or coalition billets gain an understanding of allied capabilities, limitations, and operating concepts. However, these opportunities are not afforded to all senior leaders, and the knowledge is not always recapitalized within their national service institutions.

- Should there be an effort to enhance professional education for a new generation of officers and senior noncommissioned officers, with a focus on doctrine and concepts for conducting complex, multinational amphibious operations?
- What actions are required to increase senior leader participation in NATO maritime and amphibious events—seminars, workshops, wargames, and exercises?

Personnel

Staff Capabilities and Capacity for Battlespace Management

There was disagreement among ALES participants concerning the staff capacity to facilitate C2 of a centralized ATF. When expressing preferences for force allocation and command relationships, ALES participants often made implicit assumptions about the proficiency of a given command staff to manage a complex battlespace, exercise control of specific assets (such as submarines), or conduct integrated processes such as dynamic targeting. Some participants believed that a single staff could manage all domains and friendly assets within their AO while properly synchronizing and deconflicting effects. Others believed this expectation was unrealistic given current staffs' capabilities, training opportunities, national caveats, and the authorities required.

- What are the limitations of current ATG staffs' span of control for managing multidomain operations?
- What changes in staff organization, personnel assignments, and authorities would be required to create an ATF or ATG staff capable of managing complex battlespace?
- How can current staffs' capabilities and capacity be improved?

Liaison Officers

The importance of liaison officers was highlighted throughout ALES events, with requirements identified for such officers serving in a variety of roles and capacities. For example, participants noted the need for liaisons to assist with ATF-to-ATG coordination and providing the CATF and CLF with advice regarding proper employment of national and bilateral ATGs. Other discussions centered on the coordination of specialized functions such as information warfare or cybertechnology between the ATF, JFC, and national or bilateral elements.

- What are the current and required numbers of personnel and liaison officers at NATO maritime C2 elements from each NATO amphibious ally?
- What is the appropriate number and rank of liaison officers required for a multinational ATF staff?
- How should liaison officers be selected and trained to provide the necessary synchronization and advisory capabilities?

Facilities

Command Ships

Discussions during ALES events necessarily included consideration of command platforms. Amphibious and command ships are high-demand, low-density components of any envisioned C2 construct and must be accounted for when considering the viability of any particular maritime or amphibious architecture.

- Under what circumstances could the USS *Mount Whitney* serve as the ATF and/or CFMCC flagship?
- Which American or allied amphibious ships are properly configured to serve as ATF flagships with a colocated CATF/CLF for an operation on the scale of an MJO+?
- What enhancements are required to enable these ships to function in this capacity?
- How can the Alliance better use partner ships' C2 capabilities and modifications to support maritime command staffs?

- Do allied maritime and amphibious plans have sufficient C2 flexibility and redundancy with respect to command ship allocation?
- Are there ATF C2 functions that can be exercised from shore facilities, reducing the requirement for C2 capacity afloat?

Interoperability

The Common Operating Picture and Common Intelligence Picture

Establishing a common operating picture/common intelligence picture (COP/CIP) among any large force is a challenge under the best of conditions. ALES events discussed in this report purposely provided participants with a relatively uncontested information environment in order to keep the focus on operational-level C2 constructs. Many participants noted, however, that developing and sharing a COP/CIP that enables the kind of coordination, flexibility, and mutual support these constructs require may be a limiting factor that should be explored further.

- How can NATO maritime and amphibious forces achieve a COP across the sea, air, and land domains?
- Do amphibious and maritime platforms possess the necessary hardware, bandwidth, and system architecture to develop and distribute a COP/CIP?
- Are existing systems (e.g., battlefield information collection and exploitation systems) and processes (e.g., foreign disclosure officer reviews) sufficient to share information in a timely manner?
- How can the Alliance build additional capacity with the appropriate skill sets (e.g., database management) for maintaining a COP/CIP in an MJO+ scenario?

Habitual Relationships

Some NATO countries' amphibious forces have close historical and ongoing bilateral relationships (e.g., UKNLAF and SIAF/SILF). A common theme among participants was that existing habitual relationships should be maintained and leveraged in future NATO amphibious C2 constructs.

- How can existing bilateral relationships be further strengthened?
- How can NATO take better advantage of existing bilateral relationships to improve multinational interoperability?
- Should new bilateral or trilateral relationships be created among NATO forces?
- Where do opportunities exist for allied nations with niche capabilities to be best integrated into a NATO ATF structure?

Standardization

Reexamining tactical standards for NATO amphibious forces may be needed if units will be expected to prepare for operations beyond habitual relationships.

- Are current maritime and amphibious standardization agreements sufficient?

Logistics

Although the broad notion that logistics is a national responsibility is consistent with NATO doctrine, the ability to provide at least limited logistics support across national lines could enhance the interoperability and capacity of allied forces. While the notional duration of ALES wargames was insufficient to introduce logistics stressors on any of the participant cells, there were sidebar discussions about providing common classes of supply items across elements of the force, as well as sharing common logistic services such as medical support, thus indicating participants' appreciation of the need to at least consider the possibility of coordinating some aspects of logistics beyond national lines.

- What are the candidate logistics capabilities to be coordinated or shared within a NATO ATF?
- Should the ATF include an entity to identify common and/or interoperable logistic capabilities and track status of those capabilities within the force?
- What categories of logistics and classes of supply offer greater levels of interchangeability?

Comparison of Allied Maritime Headquarters

This appendix identifies most likely candidates for the role of a CFMCC for a scenario similar to the one examined in ALES wargaming. We exclude discussion of JFC Norfolk, given its nascent state and the limited public domain information available about the command's future role. Game participants, particularly in Naples, discussed three options for sourcing a maritime headquarters in a collective defense scenario of the size and complexity explored in the TTX.[1] The below descriptions of these commands is accurate as of mid-2018. Ongoing NATO Command Structure Adaptation is likely to influence organizational missions and tasks.

Allied Maritime Command

MARCOM in Northwood, United Kingdom, is the central command for all NATO maritime forces and its commander is the maritime adviser to the Alliance. As is the case with its air and land counterparts (AIRCOM and Allied Land Command [LANDCOM]), MARCOM is a tactical command that provides advice to operational JFCs but answers directly to SACEUR.[2] MARCOM is commanded by a British three-star admiral. MARCOM is ready to command a small maritime joint operation or act as the maritime component in support of an operation short of MJO+.

Naval Striking and Support Forces NATO

SFN, in Lisbon, Portugal, is a rapidly deployable maritime headquarters that provides scalable C2 across the full spectrum of Alliance fundamental security tasks. SFN is

---

[1]   The source for information in this section is materials provided by MARFOREUR/AF, Staff discussion, Brussels, April 2017.

[2]   NATO describes MARCOM (and its adjacent commands AIRCOM and LANDCOM) as "tactical" organizations in that they oversee readiness and allocation of ships and other naval force elements; in this sense, MARCOM could also be characterized as a "functional" command.

commanded by the U.S. three-star admiral stationed in Naples, Italy, who concurrently serves as the U.S. 6th Fleet commander. The organization is managed through a memorandum of understanding between 11 NATO states. SFN is the integration mechanism for the U.S. Marine Corps, the U.S. Navy, and NATO.

High Readiness Forces (Maritime)

There are four HRF(M) headquarters that can command and control assigned forces up to the level of a NATO naval task force made up of dozens of warships. Each headquarters provides the maritime component command for the NATO Response Force on a rotational basis. One of the four organizations below is designated as the HRF(M) headquarters at any given time.

1.   Headquarters Commander French Maritime Forces
2.   Headquarters Commander Italian Maritime Forces
3.   Headquarters Commander Spanish Maritime Forces
4.   Headquarters Commander United Kingdom Maritime Forces

Any amphibious C2 construct, including the centralized ATF explored in this report, will need to "plug in" under one of these potential CFMCC options.

References

CJOS COE—*See* Combined Joint Operations from the Sea Center of Excellence.

Combined Joint Operations from the Sea Center of Excellence, presentation, Northwood, United Kingdom, November 21, 2017.

European Amphibious Initiative, presentation at the Amphibious Leaders Expeditionary Symposium, Stuttgart, Germany, October 2016.

IHS Markit, *Jane's*, online database, undated. As of June 10, 2017:
http://janes.ihs.com/DefenceEquipment/Home?tab=ALL,SEA

International Institute for Strategic Studies, *The Military Balance 2017*, online database, 2017. As of January 31, 2019:
https://www.iiss.org/publications/the-military-balance/the-military-balance-2017

Jordan, Robert S., *Alliance Strategy and Navies*, London: Pinter Publishers, 1990.

MARFOREUR/AF—*See* U.S. Marine Corps Forces Europe and Africa.

Mehta, Aaron, "NATO Weighs New Commands as Broader Reorganization Effort," *Defense News*, October 25, 2017. As of January 10, 2018:
https://www.defensenews.com/global/europe/2017/10/25/nato-weighs-new-commands-as-broader-reorganization-effort/

NATO—*See* North Atlantic Treaty Organization.

North Atlantic Treaty Organization, *Strategic Concept for the Defence and Security of the Members of the North Atlantic Treaty Organization, Adopted by Heads of State and Government at the NATO Summit in Lisbon*, Brussels: North Atlantic Treaty Organization, November 19–20, 2010. As of September 26, 2018:
https://www.nato.int/lisbon2010/strategic-concept-2010-eng.pdf

———, "Warsaw Summit Communiqué," July 9, 2016a. As of January 10, 2018:
http://www.nato.int/cps/en/natohq/official_texts_133169.htm?selectedLocale=en

———, *Allied Joint Doctrine for Maritime Operations*, AJP-3.1, Edition A, Version 1, Brussels: NATO Standardization Office, December 2016b.

———, "Brussels Summit Declaration," July 11, 2018. As of August 18, 2018:
https://www.nato.int/cps/en/natohq/official_texts_156624.htm

Ringsmose, Jens, and Sten Rynning, "Now for the Hard Part: NATO's Strategic Adaptation to Russia," *Survival*, Vol. 59, No. 3, June–July 2017, pp. 129–146.

Stoltenberg, Jens, "Press Conference," Brussels: North Atlantic Treaty Organization, June 7, 2018. As of August 21, 2018:
https://www.nato.int/cps/en/natohq/opinions_155264.htm

U.S. Marine Corps Forces Europe and Africa, "ALES 2016 After Action Report," Stuttgart, Germany: U.S. Marine Corps Forces Europe and Africa, 2016.

———, "Roadmap for Maritime Expeditionary Operations: A Multinational Approach in Europe," in *Campaign Plan 2016–2020*, 2017, not available to the public.

———, Staff discussion, Brussels, April 2017, and Stuttgart, May 2017.

———, Presentation at NATO amphibious plans and posture workshop, Arlington, Va., January 2018.

U.S. Navy Office of Information, "CNO Announces Establishment of U.S. 2nd Fleet," NNS180504-15, Washington, D.C.: U.S. Navy, May 4, 2018. As of August 24, 2018:
https://www.navy.mil/submit/display.asp?story_id=105453

Walton, Donald, "Dawn Blitz 2013: Training for Strength," NNS130128-01, Washington, D.C.: U.S. Navy, January 28, 2013. As of September 26, 2018:
https://www.navy.mil/submit/display.asp?story_id=71675